MY LIFE
IN THE RED ARMY

★ ★ ★

FRED VIRSKI

MY LIFE IN THE RED ARMY

FRED VIRSKI

My Life in the Red Army was originally published in 1949 by the Macmillan Company, New York.

Printed in the United States of America

UNCOMMON VALOR SERIES EDITION
May 2023

ISBN: 978-1951682842

About the Author

FRED VIRSKI was born in 1919 in Cracow, Poland, where he graduated from grammar school and the "gimnazjum" (secondary school). He studied chemistry at the Yagellonian University, but the war interrupted his studies in 1939. He spent five years in the Polish Army, which he joined after leaving the Russian Army, and held the rank of Second Lieutenant in a tank unit. After the end of the war, while still in the ranks of the Polish Army, he attended the University of Rome, Italy, where he almost completed his work on a master's degree in chemistry.

After his discharge from the Polish Army in 1946, he came to the United States and has since been engaged in journalism and free lance writing. His favorite activities are reading good books and listening to music—he is an enthusiastic opera lover. His favorite sports are skiing, sailing and swimming.

Author's Note

In order to protect the security of persons mentioned in this book, and particularly those among them who were my friends, I have sometimes changed their names and the names of places, as well as the numbers of regiments and divisions. These are the only liberties I have deliberately taken with the facts.

Contents

PART ONE ★ 7

PART TWO ★ 91

FRED VIRSKI

PART ONE

THE day was so foggy that standing at the trolley-car station I couldn't read the red letters of the posters on the opposite corner of the street. In the trolley, I heard rumors that a regular draft had been announced for three age groups—nineteen, twenty and twenty-one. (Mine was the last.) I was too sleepy to think about it, and too worried about getting to the factory by six o'clock. At that time being a few minutes late could mean a thirty per cent reduction of salary for three to twelve months.

This was 1940, and today was the beginning of September. I had exceptional luck with my trolley connections that day. On Waly Hetmanskie Boulevard I immediately caught my car and, luxuriously sitting on the bumper, reached my factory on Zamarstynov Street. The night watchman, old Nicholas, was sitting at the control switchboard (a Soviet innovation) in the janitor's booth. Although the clock indicated only twenty minutes to six, most of our metal tags were already hanging on their hooks. This switchboard had a glass window which could be locked, and at six old Nicholas had the duty of locking and lead-sealing it. The old man used to get so excited when the clock pointed to six and some tags were still on the table, that sometimes he would risk the grave offense of not sealing the board until 6:10; or, even worse, he would put the tag in its place on the hook, figuring that a given worker would be only a few minutes late and that perhaps he would be able to slip in unseen, using the side entrance. It was old Nicholas's luck never to be unsuccessful in this procedure.

Right after me came Puzer, who had once been the loader on the truck I drove.

"Well, Virski," roared his bass voice, "you'll be a soldier and sing 'Katiusha' [a popular Russian military song] to us on the streets."

"Don't you worry," said I indignantly. "Before you see me in a Bolshevik uniform, a lot of water will flow down the Peltev. After all, they cannot take me; I am a refugee and I have no passport!"

Puzer scratched his head.

"Maybe you're right, but I have a feeling that you'll be a *krasnoarmieyetz*." [A Red Army soldier.]

From the janitor's booth we went toward the courtyard. On the stairs we met the director, a twenty-two-year-old Russian boy, with his hands in his pockets and his cap on the back of his head. His pants did not reach his ankles; his dirty, creased coat hung on him like a bag. One had to admit that he was an honest man. Nearly all the other directors and commissars bought or requisitioned heaps of clothes for themselves. He belonged to those few who looked with disdain at "bourgeois" suits and with a kind of pride—worthy of a better cause—wore the rags brought from inside Russia.

"Good day, *tovarisch* director," Puzer and I greeted him, touching our caps.

"H'm," he muttered. "Virski"—he turned to me—"at ten o'clock we shall go to register the new car."

I nodded and spat, thinking that I would have to drive the damned jalopy to the city. Puzer and I separated: he to the storehouse to get instructions for the day, and I to the garage. Thus began a fresh day of "the new Soviet reality," in which we had already lived a year.

A week later the blanks from the drafting commissions arrived. I found out that in the spring of 1940 that they had taken the age groups 1909-10-11 of the Polish Army reserve for a kind of training. People with even one member of their family "on the other side"—i.e. under German occupation—were not accepted for this training, as they were considered "unreliable" or simply suspect. I, therefore, as a native of Cracow, with my family dispersed all over the world, should not have been subject to the draft. Unlike almost all Lwów residents, who had been forced to apply for papers, I did not have a Soviet passport. The exceptions, those who had been denied passports, were hauled off to prison and deported inside

Russia, some to Siberia, some to Kazakhstan or some other hell on earth. To this last category I, as a nonresident of Lwów, belonged. But thanks to the director of the factory, I had managed to avoid deportation. This man had given me a certificate stating that I was indispensable at the factory and that, even though a refugee, I could not be spared. This he did for me out of sheer good-will and gratitude for my having taught him how to drive and acquainted him with some auto mechanics. On the basis of that certificate I had received a piece of paper which generously permitted me to live and work in Lwów for one year.

With this in my pocket I laughed to myself as I filled out the fifteen forms connected with the draft. Letting my fantasy loose, I wrote such nonsense in answer to the question: "What is your relation to the former authorities of feudal Poland?" and to: "Do you understand that only communism can give you a truly free and happy life?" that after rereading my composition I was delighted with myself.

Under the column: "Members of your Family," I dispersed all mine so thoroughly that it looked like a mockery; yet it was the truth. My mother was in Cracow, my brother in the Polish Army in France (actually, in a German prison camp where he had been taken after the fall of France); two aunts were in Shanghai, an uncle in the U.S.A, another uncle in England, and so on. At noon I handed these papers to the director. He looked them over and scornfully puffed up his lower lip.

"And why do you enumerate all your relatives who live in the capitalist world? Do you want to boast?"

I answered innocently that I did not, that I wanted to be frank so that nobody would ever be able to reproach me for trying to conceal anything. The director looked at me with distrust; he had known me for a couple of months and realized that my *blagonadiozhnost* [Czarist term meaning "reliability"] was by no means perfect.

Though the conscription date was not yet officially known, it had already caused general excitement. The draft was to affect 150 to 200 thousand men from the entire terrain occupied by the Bolsheviks. There could not be any question of running away or not presenting oneself at the conscription center because one's family would instantly be arrested. We heard of some madmen who, if drafted, were planning to flee to the German side, where, during the fall of 1940, news was not so black. The general public, however, eyed the prospective conscription with resignation, expecting it every day.

★ ★ ★

ON the twenty-ninth of September—with exceptional efficiency, considering Soviet methods—all those subject to the draft received a notice stating when and where they should report to their draft-board commission. The first shift was called for the following day. I was in it.

In the morning I went to the factory, took my jalopy, and arrived at eight o'clock in front of the appointed building. It was the luxuriously furnished former Rail workers' Home. At the desk the first "authority" ordered me to the barber in the cellar to have my hair shaved off. Naturally I objected at once, saying that I was surely not going to be inducted to defend the Soviet Union and that therefore the loss of my hair would be a useless sorrow. My opponent was stubborn. Finally a Russian officer, hearing our yelling, came in and asked me in extra polite Polish what was going on. I explained that I was here by mistake and that just because of that mistake I had no intention of having my hair shaved off. The lieutenant listened, nodded with understanding, examined my strange passport, and decided that I would not have to be shaved, because that would be done anyway before our transportation, in a couple of days. I assured him daringly that I would not be on that transport.

Then began a comedy called "commission." In Adam's clothes I wandered from doctor to doctor. They were all Polish and you could see with what grief they were going about their business. The first one I approached was an acquaintance.

"You here?" he asked. "Say in a loud voice that you have a pain in your kidneys," he whispered.

Then in an official tone he asked me the stereotyped question, "Are you in good health?" I inhaled a great deal in my lungs and roared:

"I have a terrible pain in my kidneys, *tovarisch* doctor!"

All eyes in the room turned in my direction; almost everyone was smiling stupidly. The doctor was very seriously pressing my stomach. Not knowing where I was supposed to hurt, I screamed all the time, like a dying bull. At last the good man wrote something on my piece of paper and sent me away.

I remembered having read a humorous story of Imperial Austria, in which the conscriptees had pretended to be ill to avoid service. When I approached the laryngologist I was so deaf I couldn't hear his questions;

at the orthopedist's I pretended I had flat feet; at the oculist's, that I could not read the biggest letter on the chart. The oculist patted my back, nodded and, in a singing Wilno accent, said confidentially:

"Of course, brother, this won't help any, but try hard."

At last I reached the main room where, behind a long table covered with red felt, the People's Commission was seated. It was the strangest conglomeration, for, according to the Bolshevik recipe, it was to be a true representation of the working people. The chairman was a Russian colonel. Next him sat, poor thing, one of the professors of the University of Lwów, an old man with gray hair, representing the world of learning. There were also representatives of many other professions, among them Poles, Ukrainians, Polish Jews. On the left of the chairman sat a very young boy, a poet of the "Young Ukraine" movement, as I found out, who also had the job of translator. Almost no one knew Russian; some were slightly acquainted with Ukrainian. At that time I had twenty words of Russian and as many of Ukrainian.

The room was cold even though the day was warm. As I stood in line before the main table, my teeth chattered, and in order to pass the time I counted the pimples on the back of my predecessor. The boy had one leg shorter than the other; otherwise he was all right. I did not envy him the pimples on his back, but I would have given a lot for his leg. During the examination everyone looked enviously at his infirmity. He smiled stupidly, patting his "guarantee" of nonmilitary status, showing us how highly he valued his legs. I listened carefully when they started to question him. Behind the Commission table there was a loud deliberation; he was asked to march across the room. The colonel then asked him through the interpreter whether walking was difficult for him and if he felt well otherwise. The idiot answered that he found walking quite easy and that he was healthy as few people are. His father was a policeman; he himself had a job in the post office.

"*Kharasho!*" agreed the colonel. "In the army you will also work in the post office!"

The wretched boy was still grinning, not knowing whether he should be happy or sad. Then he limped off in the direction of the exit.

Then my turn came. All the information they needed lay in front of them on the table.

"Why is your mother in Cracow?"

"Because," said I, "she has lived there for fifty years and she did not flee in September."

"And why didn't you bring her here?" asked the colonel, as though it were possible to bring people back and forth across the frontier.

"And what for?" I asked him stupidly.

Luckily for me, he did not catch the irony of that question; most likely he thought that, according to the rules of Bolshevik education, I did not nourish any feeling toward my mother.

"Why don't you have a passport?" the chairman continued.

"Because they didn't give me one," I answered in all sincerity.

"Why not?"

"About that, I am afraid you will have to inquire at my local passport department."

"You, a student of the university, don't know the Russian language!" he went on suspiciously.

"*Tovarisch* poet," I burst out finally, "please tell *tovarisch* chairman that at our universities other languages than Russian were taught."

"The colonel," translated the poet, "wants to know whether you know another language in addition to Polish."

In one breath I enumerated eight European languages, though I knew only three. I saw the old professor cover his face with his hands to stop himself laughing.

"Sheer indecency!" growled the colonel. "He knows eight languages, but not Russian!"

Then came questions about my relatives abroad. I began telling the most incredible stories about each one, even about my uncle in New York, whom I had never seen. I tried to present each of them as a hundred per cent capitalist, and to emphasize my attachment to these relatives so as to make myself appear more suspicious. I was proud of myself, thinking that even without one shorter leg I could talk myself out of the army. After all, by now they must have been convinced of my *nieblagonadiozhnost* (nonreliability), which would make it impossible for them to take me.

Behind the table whispered discussions continued. The old professor took no part in them; from time to time he secretly smiled at me. At last the interpreter informed me that the Commission had to deliberate my case a little longer and that I should step aside and wait. I walked away and leaned against a pseudo-marble wall. Now that the nervous tension had passed, I again began to shiver with cold. But I was so sure they would reject me that I did not let that worry me. At worst, they would inform the factory that I was not "reliable," and in the factory I could manage.

After a while they asked me to appear again before the colonel. After fumbling among my papers for a few long minutes, he asked me straight:

"You're what? A driver?"

"*Da,*" said I in Russian.

"Good," said he. "You will be a driver with us in the artillery." This I understood without the interpreter's help, but at first I could not quite embrace the meaning of it. The poet-interpreter was grinning at me.

"What am I to understand by this?" I asked him. "That I am drafted?" '

"Nothing more or less," he answered.

As the next candidate took his place before the colonel, the poet turned to me.

"Nothing could have helped you," he whispered. "They take drivers without any restrictions."

Bewildered, still not realizing what had happened, I walked slowly away to the hall where our clothes were hanging. I dressed automatically, trying to realize that I had lost the battle, that there was nothing more I could do.

"Well, comrade, how is it?" somebody asked in Polish with a Russian accent. "Are we going to shave our hair or not?"

I did not have to raise my head to know that it was the lieutenant who had permitted me not to shave in the morning.

"Don't try to get fresh with me!" I said. "You might get a black eye!"

I walked slowly toward the exit. The witty one, unable to grasp my insolence, hadn't moved. At the door I turned around. There he was, staring at me with the stupor of a bull looking at a painted gate. I don't know why—I stuck my tongue out at him. I had to. Passing through the door, I burst out laughing. I realized that it was a nervous reaction, but I couldn't stop. I walked down the street, roaring with laughter. Groups of mothers, fathers, friends, waiting for their draftees, stared at me with envy.

"What? They rejected you?" asked an elderly man in the uniform of a Polish railroad worker with all the insignia torn off.

"Yes," I lied, unable to admit that they hadn't and yet be laughing.

I stopped laughing, climbed into my truck, switched on the motor, and trod on the accelerator with the vigor of a man starting a Bugatti at the races.

★ ★ ★

THE two days before shipping out I spent settling my social life in Lwów. Fortunately, there was not much of it. The worst thing for me was the fact that I could not say good-by to my father who was then in Sambor. I sent him a registered letter (telegram service was not yet functioning), telling him everything I could remember, saying good-by, and optimistically assuring him that in a few months I would probably be able to visit him. Otherwise, during all that time I was completely drunk, as I had to drink something to celebrate my departure with every single driver friend. At that time I used to drink quite a lot, although, as my brother used to say, not enthusiastically.

The night of October 1 will always remain a blank in my memory. I know that at "dawn"—as early as ten o'clock in the morning—I woke up on the table of my room, stiffly laid out on the fairly short piece of furniture and carefully covered with newspapers. I was completely dressed, even to the cap on my head. I noticed my two friends, Julius and Gregory, also arranged in very strange poses on my bed. Gregory was hugging a guitar from which he never separated in times of drunkenness. He played the guitar rather well and had a pleasant voice. I sat up on the table and tried to remember the circumstances in which we had come home. In vain. I could imagine the fright of my landlady, an oldish spinster who had a very good opinion of me, when she saw me come home last night. And when could that have been?

At this moment Gregory woke, suddenly sat up on the bed, opened his eyes wide in surprise, and for no reason began roaring the song "Gaudeamus Igitur." This did not last more than a minute because Julius, wakened by his screaming, opened one eye, contemplated him with terror for a while, sat up on his elbow, and slapped him on the ear. Without a word of protest, Gregory cuddled up to the guitar again, leaned his head against the wall, and fell asleep.

That little scene brought me back to consciousness. I looked at my watch; it was almost eleven. At one o'clock I was supposed to report to the *voyenkomat*—army office—on Mochnacki Street with my baggage. I got off the table slowly. I had a terrible headache, which made me think that I must have been drinking brandy. I went to the kitchen and put my head under the faucet.

At around 11:30 my girl friend arrived with Casimir. Together they packed my knapsack.

"I trust, Anna," said I with little hope, "that none of us started a brawl last night."

Anna looked at me in silent reproach.

"Apart from the fact that Julius put our janitor up a tree and then, in a frenzy, poked him for at least half an hour with his own cane and refused to let him down, nothing much happened. I won't go into details about how you, Gregory, and Casimir, threw stones at the monument of the *krasnoarmieyetz* on the Valy Boulevard and loudly threatened the regime. It was our luck that the streets were empty."

"Well, if that's all," said Gregory, now completely recovered, "there's really nothing to talk about."

I said good-by to the landlady without any particular tenderness and we went out onto the street. In my old Polish military coat, knapsack in hand, accompanied by a little group of friends with gloomy faces, my immediate prospects could not have been mistaken. On that day one could see similar scenes everywhere. From my house to the *voyenkomat*, following "the track of humiliation," as Julius called it, we visited fourteen saloons. It was 1:20 when we arrived at the locked door of the *voyenkomat*. Gregory was ringing the bell, while I was saying good-by to the others. Naturally, Anna was weeping. Through the door, which was ajar, protected by a metal chain, a soldier asked Gregory what was up. Gregory told him that a draftee had arrived. The *krasnoarmieyetz* said calmly that it was 1:20, and too late.

"Well," said Gregory apathetically, "if no, then let it be no," and turned to me with a stupidly happy face:

"Fred, let's go away. They already have a complete set here." But I was not so drunk as he and I knew that I had to report. All the punishments for not reporting or for being late were written out on the conscription card. I could not take any chances. I rang the bell once more; after a while the door opened and I saw a young man, also drafted, whose face looked familiar.

"Come in, man, quick, or there'll be trouble!" he cried.

I pulled the knapsack out of Julius's hands and jumped inside. The door slammed by itself. The last sounds of civilian life I heard were Anna's sobs. She was the only one who realized what was in store for me. The rest were too drunk.

★ 15

The hall was empty. The boy who had taken advantage of the guard's absence to open the door told me it had been locked at one sharp, that the men had been warned they were now soldiers and that those arriving late would be punished by being sent on the next transport—but all the way to the far East. Brr! The very thought of it made me tremble with fear and cold.

"Well, and where are we going?" I asked my savior.

"Nobody knows. Not even those two officers who are to command our transport for the duration of the trip."

"What confusion!" said I.

The boy's name was Jan; we seemed to have a number of friends in common and to have met somewhere before.

I handed in my conscription card to a dirty-looking man who was dressed in half civilian, half military clothes. As I found out later, he was one of the two officers who were to escort us. They were reserve officers, mobilized for "war" with Poland in September '39. Now they were demobilized and returning home.

We stayed at Mochnacki Street until four o'clock. In the meantime we became acquainted. There were two large groups of two hundred men each, one to become infantry, the other (mine) artillery. Rumor had it that we were to be shipped somewhere near Leningrad, close to the Finnish frontier. If this were true, why had we been told to supply ourselves with provisions for a mere three days? But I really didn't care where they shipped me.

Our little circle was quite amusing; there was Jasiu, a typical elegant good-for-nothing; Walter, a student of philosophy; Ludwik, mathematician; Julek, who, aged twenty-one, had only just managed to graduate from high school; Henry, whose hair did not have to be shaved off, since he was as bald as a knee, and who asserted that military drill went against his principles and that he would try for pharmaceutical work or for the sanitary corps.

"Have you any experience in that line?" Jan asked him eagerly.

"My uncle owned a pharmacy," explained Henry seriously.

"I see. And you worked for him," ventured Jan.

"Oh, no," laughed the "pharmacist."

"My uncle sold the pharmacy about twenty years ago, and died when I was five."

We agreed unanimously that the reference was quite sufficient for service as a first-rate pharmacist in the Soviet Army.

At four o'clock further farces began. Once more the commander-in-chief of the *voyenkomat* made a speech in which he said we were now subject to military law and that we should remember all the punishments threatening those who broke this law. Then, having formed us into groups of forty to fifty men, they opened the door and out marched the divisions of the future "defenders of the Soviet Union." We were heading toward the baths. Now these baths, as I was to find out later, are an extremely important institution in the Russian Army. During garrison service, for instance, they meant quitting camp and going to town to do some shopping. The baths were connected with the so-called *deskamera*—disinfecting chamber. This chamber consisted of a primitive vapor stove where one threw one's clothes and uniform, and where, at a very high temperature, all lice were supposed to be exterminated. Unfortunately, during the war, in Russia, where they were really indispensable, these "chambers" did not always function properly, so that most of the time the lice which devoured us were merely bewildered and put to sleep by this procedure. When they regained consciousness they threw themselves into the attack with doubled fury. Here in Lwów, however, none of us had lice; in fact, we didn't know what a louse looked like. The manager of the baths warned us that any leather article put into the chamber would burn or at least shrink. Under the shower I suddenly remembered that I had given them my cap, which had a leather strap on the inside. Jan had done the same thing. After these caps had been through the stove they might have fitted five-year-old children. Neither Jan nor I could squeeze them on our heads. Finally Jasiu helped us by cutting the burnt straps, squeezing the sad remainders of the caps on our heads, then tying them with a string under our chins.

At every opportunity—before leaving the *voyenkomat*, before entering the baths, after leaving them—the authorities took great care to count us. I had the feeling that the officers doing the counting were each time surprised to see that no one was missing. This procedure I considered very stupid. Desertion was virtually impossible because in the Bolshevik system the family was responsible for all crimes committed by any of its members. Surely no one was going to report to his draft board in order to escape from the baths!

At about seven o'clock, when all groups had finished their washing, we were formed into columns of four and marched through the city to the Railworker's Home, where exercises, performances, and functions were about to take place. We had covered a few meters when the lieutenant who was leading us cried, "Let's sing, boys!" and he began the famous "Katiusha." Not more than a dozen voices harmonized timidly with his; the rest, more than two hundred men, kept a dead silence. The lieutenant couldn't understand why we didn't all want to sing. Weren't we going into the army, where we'd be given food and clothes, and not have to work like horses for a piece of bread! Nobody tried to explain to him. We had already reached Akademicka Street, where mobs of sad-faced people were standing around on the sidewalks. We marched in silence. All of a sudden Julek began to hum an old Polish military song. *"Jak to na wojence nee ladnie"* (How pretty is war!) By the song's fifth bar two hundred voices were roaring its refrain. Beside me marched a young Ukrainian, howling "When the Uhlan falls off his horse" (one of the song's lines) with such vigor that the veins of his neck swelled. I gave him a friendly smile. Without interrupting his singing, he winked at me, smiled back, and roared even more forcefully. The people on the sidewalks were smiling between their tears. When an old woman at the corner of Akademicka and Legionow streets blessed us with the sign of the cross, somebody shouted to her from the ranks:

"Don't cry, mother, we'll be back soon!"

I remembered all the conscriptions I had seen, read or heard about. The soldiers always marched gaily away, singing, the mobs accompanying them with flowers. What I observed now looked like a funeral procession, in spite of the songs. After a few stanzas the lieutenant screamed "Enough!" from the head of the column, but nobody paid any attention to him. Finally, exhausted, we had to stop and march in silence. We all had the childish feeling that we had played a good trick.

In the Railworkers' Home, a military band was playing to welcome us. There was also a buffet, free for us drafted men and cheap for the guests, as it was governmental. In order to get free tea I had to show my papers, as my head was not shaved; with my hair, no one believed I'd been drafted. I was the only one of four hundred men who still had his hair.

Anna, Julius, and Gregory had been waiting for me for a long time. They all looked just the same as when I left them: Anna's eyes were swollen; Julius and Gregory were still thoroughly drunk. We had little to say to each other; there was nothing to talk about. Nobody had any plans; to speak of the future was aimless. After a short silence we went to the bar and for at least an hour drank *moskovskaya* [45% proof vodka] with resignation. To our terror, even Anna drank two glasses. Then our tongues became more flexible and we could chat without order or association. Gregory told silly jokes; Julius repeated with a drunkard's stubbornness that the war must be over in a few months; Anna swallowed her tears and smiled meaninglessly. I thought of my father, who in two or three days would receive my letter informing him that I'd been drafted into the Russian Army, by which time I would no longer be in Lwów.

At eleven o'clock we were ordered out into the street with our baggage. Mothers, wives, sisters were sobbing. The hour of farewell had arrived. *Moskovskaya* was singing in my head; I took my coat and threw my knapsack over my shoulder.

"Friends," said I, turning to my little group, "let us not despair. Let's grin and bear it, say good-by, and hope we shall meet again."

"All right, Fred, I promise you I won't cry!" said Anna in a broken voice, while I hugged her tight. "I know you're talking to me and not to the boys."

"Thank you, Anna, for all the good and bad moments that you bore so bravely with me. Try to take care of my father, and write." What else could I say to her? That I would return, that she should wait for me or something equally stupid? I felt then that I would never see these friends or Lwów again. Julius embraced me without a word; Gregory wetted me in his drunken manner, and between loud hiccups swore that from now on, as a sign of mourning, he would never touch vodka again. (A month later, Anna wrote me that Gregory's life was in danger from alcoholic poisoning.)

As Jan and I took our places in the ranks, Julius shouted that my good friends, Casimir and Taddeus, would probably be waiting for me at the station. The band started at the head of the procession. On the sidewalks ran little groups of families, anxious to be with their boys to the last moment. After a few meters, Anna rushed up to me again, but this time her crying resounded to high heaven. We walked together for a while, then I kissed her once more and pushed her back to the sidewalk straight into the arms of Julius.

"Please go home now," I begged him. "It is after eleven. Anna should be in bed."

Julius nodded his head; he was holding Anna, who was trying to tear herself from his hands. As I was marching with my head turned around, I tripped and would have fallen if Jan had not caught me by the elbow. I turned again and saw them standing on the same spot. Julius and Gregory, holding the poor girl, raised their caps in a last farewell. I raised my arm once more and turned violently away.

Jan smiled sadly.

"Damn it all, one just has to forget all these people. I left my entire family and fiancee behind." He also felt that this farewell was for neither one year nor two.

A long line of freight cars stood at the main railroad station, half of which had been destroyed by the Germans in 1939.

"Forty men to a car!" yelled the officers. Each car contained two long double-decker platforms for sleeping. Julek was the first to get on board. When our turn came, he had occupied one whole lower berth and admitted to it only the members of the newly organized group of friends. We put our bags at the heads of the berths, covered the boards with coats or sweaters, then lighted a stable lamp which hung from the ceiling. So far, very few people had managed to get onto the tracks to see the train off. I looked out of the window for my two friends, Casimir and Taddeus, but in vain. At about one o'clock they added to our transport ten cars of drafted men from the regions of Janow and Gródek Jagiellonski. The men in these cars were quite different. The element in our transport was entirely urban, and though mixed—Poles, Ukrainians, and Jews—it had a unified standard. And nine out of ten spoke Polish. The other transport was made up of peasants' sons, mostly Ukrainian plowboys taken straight from the field or the stable. As they moved along the tracks you could hear mandolins, balalaykas, and accordions playing sentimental Ukrainian songs which to me always sounded like mournful and plaintive howling. Jasiu, who had been to a Ukrainian high school, cried to them in the darkness: "Where you from, boys?"

"Janov, Trembovla, sir," replied the sad and frightened voices. When the "country" had been united to the "city," as Walter expressed it, all along the track people began shouting and waving lanterns—which in Russia is a sign that in about half an hour a train may begin to move. Just before the engine whistled, I heard cursing at the track entrance and then the voice of Casimir, yelling to high heaven:

"Fred, Fred, help! These sons of bitches won't let me on!" Taddeus was whistling our signal, the first bars of Rimski-Korsakov's "Caucasian Suite."

"Here!" I yelled, "right across from the entrance."

After a while Casimir appeared alone in the darkness and handed me a heavy cardboard box.

"Fred, we couldn't find you anywhere; we didn't know about that celebration at the Railworkers' Home. Then we were misinformed and waited three hours at the freight depot. Now they won't let Taddeus on the track."

The locomotive whistled once more.

"Thank you, Casimir. Keep well," said I.

Then the engine gave a great snort and, asthmatically pulling, slowly began to move.

Casimir ran along beside the train.

"Don't give up, Fred. Everything will be all right!"

The engine was gaining speed, and Casimir growing smaller and smaller, till I at last lost sight of him behind the station.

The city was poorly lighted; I could still see a light on the citadel and another way up on the Wysoki Zamek. Then everything disappeared, and we plowed into the darkness. All forty of us milled around the two doors of the car. No one spoke. For all of us Lwów was the symbol of our young lives, our homes and families. Even I, a Cracovian who had lived in Lwów for only a year, loved that city dearly. Three hundred and fifty thousand before the war, it was today a city of one million inhabitants, pulsating with life, wit, and music. From the tail of our transport the wind brought the strains of a plaintive melody.

"Damn them all!" cursed Julek. "Are they going to howl like this all night?"

That was the end of daydreaming. We began to disperse to our berths. The lamp in the ceiling died. I spread out on my space and lit a cigarette. From my corner I could see the two opposite berths. From both, the tips of lighted cigarettes glowed. Nobody was asleep. The train was ruthlessly shaking. Straw and clay kept dropping from the upper berth. I placed my knapsack in such a way that I could lean my head on it and at the same time use it as a protective roof above me.

I began to think, though I tried not to. A new period was beginning in my life as well as in the lives of my companions; but for them separation from their families had only just begun, while I had been without a home for a year. My position was that much simpler, for while they were now

★ 21

losing the comforts of their homes, their greater or smaller possessions, I had lost all those things in September '39. I now had nothing material to lose. My earthly possessions were on me or in my knapsack. A year ago I'd been a university student. After that, a driver. I had not sneered at my new profession; on the contrary, I blessed it. But for it, I don't know how my father and I would have managed. After all, the jump from student to driver had been so big that my present jump from driver to *krasnoarmieyetz* was nothing in comparison. I thought of Anna. A kind girl, to whom I owed much. Her home was always an oasis for me, where at any time of the day or night I found the sincere friendliness of her parents, particularly her father, an engineer, with whom I would often take a glass of vodka, and Anna's own quiet but attentive love. Did I love her? I didn't know, but probably not. I was attached and so accustomed to her that sometimes I had the feeling I had known her for centuries, even though it had been only a year. When I visited her in the evenings, I was so exhausted after a day's work that without a word I'd spread out on the couch and fall asleep. Then the entire family would walk on their toes, while Anna sat by my side. I was now reminded of what we had gone through together. At first, when my father was ill at home and I worked eighteen hours a day, she would come during my absence to clean the room, cook something for Father, or make a fire in the stove. I thought of the time when she and her family (her father was a reserve officer in the Polish Army) were to be deported, and later when the militia were searching for my father and myself. And I thought of the time, the only time, I betrayed her stupidly. I have no idea why, but probably out of spite, I began to see her girl friend. Three weeks later I went back to Anna with repentance. The day I returned she received me as though she had seen me an hour before. Such was Anna. Before my departure I had explained to her that, of course, we could not count on each other and that we should both consider ourselves quite unbound. Though Anna nodded, I knew that she did not agree with me.

The other problem which did not let me sleep that night in the unbearably shaky cattle train which was taking me to an unknown destination, was of a political nature. It was clear to me that the war would not be over soon. It was also clear that sooner or later Russia would enter the war. The military potentialities of Russia, stored for years, would have to be discharged. Today Russia was an ally of Germany. If she entered the war on the side of Germany, against the Allies, I would be forced to fight everything that had been dear to me for years, on the side of the two greatest foes of my country. If Russia were to go against Germany (or vice versa), which was generally considered possible, I would gladly fight against the Germans, but for whose sake? In whose interest? Russia's?

The train kept stopping every fifteen minutes, pulling up sharply, then moving on again. Now it stopped at an illuminated station.

I got down from the berth and looked through the half-closed door. Tarnopol, a city I knew as well as my own hand. The platform was empty. Should I run away? Nonsense. I wouldn't be able to stay hidden for long; also, they would arrest my father within twenty-four hours. I lit a cigarette. Jan was beside me.

"I bet you were thinking of escaping," he whispered. I nodded, but waved my hand in resignation.

"Of course," agreed Jan. "There would be no sense in that." We climbed back to our berths. I put my elbow under my head and fell asleep. The train kept on plowing east.

★ ★ ★

THE sun was quite high when I woke. The train was standing in the middle of a large field. Jan, brushing his teeth in the doorway, told me we were fifteen kilometers outside Tamopol. Apparently we were waiting for some more cars of cannon fodder. I looked out of the door. The track ran beside a brook in which forty or fifty boys were washing. Far away at the head of the train the engine was hissing and throwing clouds of steam in the air. Along the bank a file of squatting "heroes" of the Soviet Union, in not quite heroic attitudes, were satisfying their major physiological needs.

"Well, army life has begun," I thought.

After our morning toilet we ate a collective breakfast under Jan's leadership. Jan was probably the most sensible boy in our group. He had excellent ideas about making life simple, but when it came to putting them into practice he was a bit of a ninny. After a year in the army he could not sew a button on his trousers. Now he was dividing up the work and the food. When I returned with a pail of boiling water from the engine-driver for our tea, Jasiu was smearing bread with lard, while Walter was "setting the table"—covering two boards with newspapers, then placing them so that we could reach the food from two sides. Opening the box Casimir had brought me the previous night, I found a few kilograms of dry sausage, sugar, cacao, cigarettes, and four bottles of liquor. Three of them were some ordinary "postwar" brand, but one was original *wyborowa czysta*—the best Polish vodka. The discovery caused general applause among the gang and we crowned the breakfast, rather unusually, with slivovitz and sausage. Around noon four cars were added to our train. Jasiu went to investigate. He came back waving his hand in resignation.

"No, nobody from Lwów," he groaned. "They are all Huculs." [Polish-Ukrainian peasants from the eastern part of the Carpathians.]

We moved on, stopping at even the smallest stations. At about four o'clock we reached Podvolotchyska. Behind the depot, a decent stone building, ran the river Zbrütch. On the other side of the Zbrütch was Russia. We had stopped in such a strange position that the locomotive was on the bridge, while our car was almost on the bank of the river. Next to the bridge a destroyed frontier post with an eagle and the letter "P" was lying on the ground. It lay there, a symbol of everything that had fallen more than a year ago. Through here ran the now nonexistent boundary between Russia and the now nonexistent Polish state.

"If that post were standing up the way it was a year ago," remarked Jasiu, "we would not be going to wherever we are going!"

We all burst out laughing.

"Jasiu," decided Walter, "that sentence should be engraved in golden letters on marble. It is the most accurate answer to the question as to why we are going to Russia. Because our frontier post fell down!" He was choking with laughter.

The engine whistled and the train rolled slowly over the bridge, then stopped at the station of Volotchyska, already on Russian territory. The building was made of wood and was extremely neglected. The paths leading to it were covered with deep mud. In front of the building we noticed a woman in uniform, high boots, and the red cap of a stationmaster. She was taking tremendous, manly strides, beating her high boots with a little folded flag. We had seen a great number of Russian women in Lwów, but this one in her uniform, boots, and man's cap was a revelation to us.

"So that's a woman!" cracked Dziunek.

"Gentlemen," added someone else, "I'm switching to little boys."

The general impression of the first Russian town was rather morbid.

"What surprises me," remarked Walter, "is the fact that the Bolsheviks, specialists in propaganda, did not build an exemplary station to make a good impression on visitors."

"Man," retorted Jan, "for whom would they do that? Tourists were not permitted here, and the first 'foreign' guests to cross the frontier at this point were probably our own people being deported inside Russia. Now it's ourselves, the army of His Dictatorial Excellency—Stalin! I suppose you thought they'd construct a triumphal arch to greet you!"

A few youths in railroad uniforms, carrying shovels and picks, appeared from the station, climbed onto the bank, and in single file walked along the track. The first one, with a cigarette made of newspaper in his mouth, came up to Julek and with a gesture asked him for a light. Julek took out a pack of cigarettes, offered one to the man, then to the others. The first man hid his newspaper cigarette carefully under his cap and with great interest turned the other one in his fingers.

"Who are you?" he asked. "War prisoners or arrested people?"

"War prisoners! Arrested people!" yelled Julek. "I bet you've never seen a transport carrying anyone else!"

Jan defended the worker who, not understanding Polish, was astounded by Julek's outburst. "We are recruits," he explained politely. -

The railroad men were all very surprised. Our transport was, in fact, the first of its kind to pass their station. We had been offended by their question—not that it was offensive, but because it reminded us of the thousands of our people who had taken this route as arrested persons.

Soon we began to cover our first kilometers on Russian soil. The trip dragged unbearably. All we knew was that we were moving distinctly to the east. At about four o'clock the following afternoon we stopped at the major station of Zhmerinka. Here, in the waiting room, we found a meal prepared for us. The tables were set with tin plates filled with some kind of stewed meat with kasha. It was hot and that's what we'd been missing most.

Our gang was among the first to eat, and afterwards, against all regulations, we went out on the streets. Extreme poverty greeted us from every corner. The dress of the inhabitants was consistently poor. The only store window with anything on display was a sportswear shop. Tennis rackets and complete fishing sets hung behind cardboard fish. I should say here that after two years in Russia I never saw a tennis court, or a man fishing for the fun of it. There was a war on, of course, but that's no reason why tennis courts should have disappeared. People looked at our clothes and boots, at our faces, as if they were seeing creatures from a different world. Dziunek made a gesture of resignation.

"Let's get back to the train," he proposed sadly. "Looking at this makes one cry."

Around noon of the following day we arrived in Kotovsk. Here they took off all the cars except those containing the more or less five hundred of us from Lwów. Then we moved on. Jasiu, who had jumped into the train at the last minute, began telling us of an interview he had had with a railway employee at the station. Our direction, he said, was now decidedly south, and our destination could be only Odessa or some other town on the Black Sea.

"Why shouldn't they assign us to a garrison in some dead hole in the country?" asked Ludwik. "Why keep running on about a town?"

"They must send us to a town," Jasiu repeated stubbornly. "What would we do in the country?"

"Optimist," growled Ludwik.

We were traveling pretty fast now; the landscape was flat but nicely cultivated. Toward evening we felt certain we were heading for Odessa. Around nine o'clock, the train slowed down in a big freight station. We rushed to the door. At last, on a glass booth, we read; Odessa. The station was huge; we kept going through it, continuously changing direction. Finally we stopped beside a concrete loading platform. Opposite, men and women were loading watermelons onto freight cars. The officers in our transport were running alongside the train, yelling that no one was to move one step from the transport, that trucks would soon come and take us to the barracks.

When the officers had disappeared and the workmen finished loading and scattered, we wandered about on the dirty platform, and to kill our boredom, we ate melons. Around eleven we decided that Jasiu should go and get information. Meanwhile, some of the boys began arranging their bedding on heaps of melons. A passing railway employee happened to notice that one of them had taken off his shoes, put them aside, and wrapping himself in his coat, was getting ready to sleep. The employee stopped, poked the shoes with his stick, and said laconically, "Hey, they'll be stolen!" Seeing the astonished look on the boy's face, he added, "Hide them under your head." From that day on, every one of us knew that when it came to sleeping in the open air, shoes must be under the head.

When Jasiu returned twenty minutes later he was delighted.

"A few hundred yards from here there's an exit," he told us. "True, there is a watchman, but he'll let us through. I've already given him a few cigarettes. Beyond the exit there's a 'dancing' in the open air. To a phonograph, of course. You can get beer and wine."

We threw out things together in one big heap and went like one man. Henry, whom the vision of wine did not attract, promised to watch our belongings, because, as he claimed, for once he was not sleepy.

"When they come for us," Walter yelled to him, "throw the things on the truck and tell the colonel they made us wait too long."

The exit was there, all right. So was the watchman, who, on receiving a whole pack of cigarettes from Ludwik, opened the gate with a smile which almost tore his face in two. Beyond the gate, illuminated by multicolored lamps, was the railway "dance hall"—an open garden the size of four tennis courts, with a tall wooden fence separating it from the street. At the entrance there was a ticket booth—admission fifty kopecks. I gaped around a little, so that I was the last to buy my ticket. I put one ruble on the counter; a hand (quite pretty—in any case, feminine) began to count my change in kopecks. As the window was low, I bent down and stuck my head almost into the booth. The impression was thundering! A small head with blonde curls, a little, slightly upturned nose, nicely shaped and strongly carmined lips. Very inviting breasts daringly thrusting forth the cheap cotton dress. Blue eyes were looking at me gaily from under long lashes. It all formed a very nice picture, so much the nicer because totally unexpected.

"By what miracle has nobody discovered her yet?" was my first thought. Then I told myself that they had all been too lazy to bend down while buying their tickets. We looked at each other for a while, my change still lying on the counter. It's rather difficult to start a romance when one knows only a few words in the language of the object. At last I smiled, rather stupidly, I suppose, and said, "Good day." I couldn't think of anything cleverer. The girl answered me in the same way and continued to smile. What on earth could I say? Then she helped me by asking, "Are you people from the West?"—which was how Russians described people from that part of Poland occupied by Russia. I was very glad of this unexpected assistance. "Yes," I answered eagerly, and in broken Russian, "We are conscripts from Lwów. We arrived today."

The girl smiled broadly.

"Lwów is a nice city," she told me, which I confirmed with as much pride as if I were its mayor. Had she ever been there? No, but her father had, and had told her about it. The conversation, in spite of my very limited vocabulary, started to gather impetus. It turned out that the girl's name was Nina. She was nineteen years old, knew a little French, with which we patched the conversation, was studying in a teachers' seminary, and was making a little extra money in the evenings as ticket seller. I was charmed. Wondering what kind of figure and legs she had, I asked her if she could come out to the "dance hall." She seemed embarrassed and explained that she was not allowed to leave the cashbox. Too bad. I continued to stand by the booth with my head stuck in the window. It was terribly uncomfortable, and my back hurt; but what won't one do for an ideal?

Suddenly the door opened and Jan appeared.

"What the hell's happened to you?" he asked almost angrily, seeing my strangely twisted form and the dazed rapture in my eyes. "I've been looking for you everywhere in this luna-park, and here you are having a chat. Without a word to anyone!"

I tried very hard to look mysterious. "Jan, old man," said I, "don't be mad at me just because I saw something nice in a window."

"I don't give a damn about you or your window," growled Jan with a gloomy face. Don't pull such tricks on me. As if all I have to do is to worry about you!" Nevertheless, he fixed his tie and came nearer to the booth.

"Please note that all rights are reserved," I hastened to warn him.

Jan was dangerous competition. I beat him inasmuch as I still had a head of hair, while his had been shaved. But he was taller, sturdier, and had beautiful teeth which he displayed in a charming smile. My only consolation was that his linguistic talents were even poorer than mine. Now he stuck his huge head into the window, showed his "deadly" teeth and, after a while, turned to me.

"I forgive you," he said generously. "This one could stop even me!"

He stuck his head back into the window, smiled again, and said in Polish, "Lovely."

Nina bent down to me. "What did he say?" she asked, glad of such a rush of admirers.

"He wants to buy another ticket," I lied calmly.

Nina looked surprised. "Another? What for?"

I shrugged my shoulders. "If he wants it, give it him," said I.

Jan was listening to my explanations and the girl's questions with a charming smile. Not understanding a word, he was certain I was translating his compliment. Shrugging her shoulders, Nina put a ticket on the counter and pushed it toward Jan. He made a stupid face. The girl showed him five fingers. My change from the ruble was still lying on the counter.

"I'll treat you if you'll go away," I proposed politely, and told Nina to take the fifty kopecks. The girl put the change in the drawer. Furious, Jan took the ticket automatically and asked me what it all meant.

"Nothing," I answered with stoic calm. "After all, you're going back into the dance hall, and you have to pay admission every time. When in Rome, do as Romans do," I added.

Jan looked at me incredulously, went to the door, opened it, and then turned around laughing and called:

"You got rid of me, damn you! But it cost you fifty kopecks!"

"It was worth that much," said I, showing my teeth in a joyful smile. The door slammed behind him.

"Why was he laughing?" asked Nina curiously.

"He thinks our eyes are the same color," I invented quickly.

Nina peeped curiously under my cap.

"Indeed, yours are also blue," she announced with satisfaction.

We chatted like old friends, helping ourselves with our hands. My spine was already completely stiff. At last Nina suggested that I go through the door to a larger window which was closed. When she opened it I could see her much better and had a much bigger field of action.

Considering it was my first romance in Russian, I was doing very well. As a matter of fact, we had already told each other all that was necessary; I was making her unequivocal propositions and she, without a shadow of embarrassment, was accepting them. I was astonished; I did not as yet know the freedom of Russian women. She explained to me very convincingly that the dancing would be over in an hour and we could go to the park. I confess that her readiness spoiled my mood a little. But after all!

Ludwik and Jasiu came up to me with a young girl and a Russian boy.

"New friends," said Jasiu, introducing me. It turned out that the girl, Zhenia, was a friend of Nina's, and that her brother, Mikhail, only seventeen years old, was envious of our being in the army.

"Tell him he can exchange places with me," I told Jasiu, who, with his knowledge of Ukrainian, was our general interpreter.

"Exchange!" cried Jasiu. "D'you know what he does? He works in a factory eight hours a day, does four hours extra work somewhere, goes to school in the evening—and at this moment his whole fortune consists of one and a half rubles."

"To hell with him," I said. "I'd rather be in the army."

They then took me to an open pavilion where the whole company had gathered, civilians and girls included. Dziunek, our Croesus, was buying one bottle of wine after another, each bottle costing six rubles.[The pay of a private in the Red Army was 10 rubles a month.] Julek sat in a corner passionately kissing a brunette; even Walter was embracing two quite good-looking but horribly dressed girls. Such impossible dresses are not to be seen anywhere else in the world.

Through a megaphone somebody was yelling in Russian that "in honor of our guests from Lwów" a tango from the film "*Wesolaia Rebiata*" *(Gay Boys)* would be played. The ladies were to invite the gentlemen to dance. The voice seemed strangely familiar. The public was delighted and the dance floor began to fill. Then, from the phonograph booth, the operator's face appeared. It was Ludwik, smiling broadly. It seemed that he had been sitting there for an hour, thinking up new attractions. Coming from a small village on the Polish-Russian border, Ludwik spoke Russian as well as Polish.

At last the megaphone gave out a final melody. Ludwik took leave of the whole company. The park was getting empty. Someone was putting out all the lamps, one after the other. Nina was closing the box office. At last she opened the door and came out with me into the street. The night was so dark, not only could I not get a look at her legs, I couldn't even see the road ahead of me.

She put her hand under my arm, came closer, and said, "We'll go to the park." She had a strangely swinging walk, but I did not pay any attention to it. In the darkness Jan's voice reached me:

"Fredek, the watchman says so far nothing's happened, but don't stay away too long."

"I'll be right back," I called, and went further into the park with Nina. We found—that is, she found—a bench. It was so dark that I could hardly see the white spot of her face. . . .

Later, walking back in the direction of the railway station, she gave me her address, and I promised to write her from the barracks the very next day. Near the railroad crossing we stopped under a lonely lantern.

"Don't be angry if I don't take you home," I apologized, "but I wouldn't be able to find my way back."

"Of course," she reassured me. "It is only a few steps to where I live."

"Good-by, Nina," said I, putting out my hand.

Do svidania. Good-by," she answered, and turning, walked slowly away. I stood there, staring after her with stupefaction. Only now I understood why she had that swinging walk! One of her legs was definitely much shorter than the other.

Finally she disappeared in the darkness. I stuck my hands in my pockets and headed back to the station.

I found my gang encamped on the luggage and asleep. I lay down and had not slept an hour when a noise woke me. Along the rows of melons I saw a line of trucks, their headlights illuminating the whole area. Some of the boys were already piling in. In the middle of the square, under a lantern, stood a young noncom. "Infantry—first ten trucks! Artillery the next!" he kept yelling. Jan was furious.

"If they could get along without us till four in the morning," he groaned in a sleepy voice, "they might have let us sleep till daylight."

Julek and Ludwik were already aboard; we handed them our things and away we went. We drove through rugged suburban streets until we were almost outside the town. At the end of some trolley rails stood three huge buildings surrounded by barbed wire.

"A prison or our barracks," Jasiu predicted.

As he spoke, we drove through an immense gateway, past a sentry box, to the barracks square. The noncom led us to a large hall on the ground floor. From the ceiling shone a weak red lamp. We sat down on the floor against the walls.

When dawn broke we saw a young officer approaching from the main building. His various stripes and white armband suggested an officer on duty. He came into the hall, looked at us, and we at him. Then he asked whether anyone understood Russian. Ludwik came forward. Instructing him to take command, the officer led us out to a sports field and told us to wait until the regiment had finished eating; then he'd fetch us for our breakfast.

We had been lounging about on the field for a while when up came a little group of officers. In the middle of it walked a pleasant-looking fatty, more like the owner of a grocery store than a Red Army major. We deliberated for a while whether we should receive the officers sitting down or standing up. Jan sensibly decided that we should get up.

"Which of you is a translator?" asked the major.

Ludwik stepped out. "I am, *tovarisch* major."

The major asked Ludwik to greet us in his name and in that of the regiment and to express the hope that we would be pleased with one another. Then he suddenly asked whether there were any soccer players among us. When quite a few put up their hands, the major smiled widely with contentment.

"Go to breakfast now," he told us through Ludwik, "then we shall see how well you can play."

The mess was a tremendous hall, packed with tables and benches. In the center of the hall a noncom was rapping out commands: "Come in! Sixteen men to a table! Eight to a side!"

In front of each soldier lay an enamel bowl. Two small baskets contained bread, three slices per man. The soldier on duty planted two pots of tea in front of us. "One pot for eight men," he announced. Another threw on the table four lumps of sugar, one lump to be divided among four. Then up came another soldier with a bowl of herring, which he began throwing into our plates. His aim was very good, but the herring smelled horrible. They were freshly salted, straight from a barrel, with heads and all the insides. I am not over-delicate, but I felt uncomfortable. When the soldier had passed on to the next table with his stinking horn of abundance, Jan made a face suggesting he was going to throw up. Julek, holding his herring by the tail between two fingers, was swinging it back and forth like a pendulum.

"What are you up to?" asked Ludwik.

"I'm going to throw the herring back into the pot!" roared Julek.

"Better not," warned Ludwik. "You'll miss the pot, then there'll be a row. Anyway, you might as well know that in Russia herring is a delicacy, and that they must have wanted to please us."

Ludwik had worked in a Russian office in his home town and knew many things that were quite strange to us. Though some of the boys ate the herring, it looked as though most of us were going to have dry bread and tea for breakfast. Lolek was passing from table to table, collecting herring from those who could not stomach them.

"Do you really like them?" Dziunek asked him in surprise.

"Are you mad?" answered the fat Lolek indignantly. "If we refuse to eat, we'll simply die of hunger!"

At the next table the soldiers who had cooked our breakfast were now eating theirs. Holding the fish in one hand by the head, in the other by the tail, they ate it like corn, spitting out the bones onto the table. Before reaching their dirty paws, the herring had passed through a dozen similarly "clean little hands." Then a soldier appeared, collected the untouched herring, popped them back in the pot, and in their place gave the hungry men a piece of salted lard.

After leaving the mess, we returned to the sports field. Here, two soccer captains picked their teams, eleven to a team. Five or six boys had been members of excellent soccer teams in Lwów. Then a noncom arrived with two soccer balls. From our boys Jasiu, Julek, and, to our great surprise, Ludwik, were chosen for the teams. While the players practiced, we, the spectators, sat down under the wall and listened to the noncom telling us who was who in the group of officers that had reappeared with the major. The tall, solidly built, kind-looking owner of two rectangles on his collar was Shurin, the regiment commissar. The short, round, silly-faced man was the chief of staff, Lermontov. The tall captain with sharp, handsome features covered with veins was the engineer, Kovalski, supposedly a Pole, and certainly a scare to all the drivers and tractor men. Finally the tiny, terribly thin senior lieutenant with the dry, nervous, intelligent face was Kalugin, known as the regiment's most talented officer. Our battery was to be assigned to him.

"Well, in that case let's hope they won't separate us," sighed Jan piously.

The major was looking at the players sympathetically. Crazy about sports in general, he was a particular enthusiast for soccer. Suddenly he drew out from the depths of his breeches a treble-voiced judge's whistle, strode onto the field, deliberated with the players, and whistled for the match to begin.

How little youth needs to be happy! Here we were, all broken by the separation from our loved ones, the morbid prospect of service in a foreign army, the recent shock at the breakfast table, and yet, in spite of all that, the players were kicking the ball with the same ardor as they had in the suburbs of Lwów, and we, the spectators, sitting under the wall, were roaring just as loudly.

Everybody forgot his sorrows. The major, with a joyful, kindly face, ran about the field, whistling and shouting most enthusiastically of all. After the first few minutes, still running, he took off his rubakha [green military blouse], belt, and pistol, and tearing along the line of spectators, threw them to one of the boys. I have never known much about soccer, but at close quarters I rather liked it; there was tempo, cooperation, and intelligence on both sides. The major was now whistling for an intermission, with the score at one all. With a great roar from the spectators, the players left the field and centered around the distinctly enthused major. He stood among them with a red, perspiring face, in a jersey undershirt, giving out advice, direction, and high praise. The boys were very friendly with him. Julek, with a happy face, called to him in Polish, figuring that nobody would understand:

"Major, you're not a bad chap!"

I looked at Kovalski. A smile passed over his face; he had understood but he didn't let anybody notice. The major was in heaven. He asked about other branches of sports. Among us were apparently two high-class boxers, one the champion heavyweight, the other a vice-champion lightweight of Lwów. There were also three sprinters, two fencers, and a few oarsmen. The major was almost in tears from happiness.

"Well," he repeated in joy, "at last we are going to have a sporty life in the regiment."

The commissar was not such a sports enthusiast. Approaching the group surrounding the major, he asked with a smile:

"Well, and how many of you have read the history of the Bolshevik party?"

Dead silence was the answer. The major made a resigned movement with his hand.

"Don't worry," he told the commissar. "They can read here. That's your business." Then he looked at his watch and whistled for the second half of the match.

When it was over the score showed two all and the time almost ten o'clock. The major said good-by; we took our things and, led by a few noncoms, set off in the direction of some long, low buildings at the end of the regimental grounds. These, as we found out later, were storage houses and workshops. The last two buildings were empty. Here we were to establish ourselves for the period of our quarantine, which was to be spent getting our uniforms and having ourselves assigned to a battery.

The following day we made a tour of the storage houses. Among them we found a *larok*, or soldiers' shop. The salesman, an elderly Jew, was standing behind the counter, which was heaped with halvah and packs of *makhorka* [black, coarsely cut tobacco]. We were surprised by the halvah which, though not excellent, was at least sweet. Odessa was probably the only city in Russia possessing candy factories; halvah, as a semi-scrap product, could be found here in great quantities. This little store, moreover, was a hundred times better equipped than the civilian stores in the city. Such articles as pencils, pens, paper, needles, thread, buttons, shoe polish, as well as other accessories of military life, were quite often available. Sometimes it also carried cigarettes, of which certain brands were not at all bad, though very expensive. A pack of twenty-five Bielomor kanal, for example, cost two rubles—one-fifth of our month's pay. You could also buy lemonade, the so-called *kvass,* sometimes even beer and sweet rolls.

We, however, were not the store's only customers. The wives of *komandirs* [the general designation for superiors in the Red Army, used for both officers and noncoms] used it too. Appearing regularly every week for their monthly rations of potatoes, fats, and sugar, they were mostly young women, terribly poor, and very badly dressed, particularly in the winter. In the spring, when all they needed was a piece of cotton material, they looked much more decent.

Roaming around the terrain of the camp, we discovered only one other attraction. Along the wire fence near the kitchen there always gathered at mealtimes a crowd of children between the ages of five and twelve, each of them carrying a pail. Into these pails the soldiers who worked in the kitchen threw the remainders and scraps from our meals. For half an hour one day I watched an eight-year-old boy, filthy and in rags. Having found in his pail quite a bit of groats, or kasha, mixed with some soup and a handful of biscuits, the boy drew from his belt a wooden spoon. With this he began eating from the pail, fishing out the better pieces and leaving the rest. Then he laid out in front of him on the ground the pieces of biscuits, which he assorted attentively: the whole pieces he put aside in the torn pocket of his jacket, the crumbs he poured into his cap. Then off he went down the muddy road, along the wire fence, in the direction of the suburbs. The cook, who had been watching the scene, smiled and said:

"Well, he'll make a couple of rubles."

From a conversation with the cook, Jasiu discovered that these children begged partly for food for themselves and their families, and partly in order to sell their products to those who had no children to send to the wire fence.

"A beautiful life," said Jan, shaking his head, " 'under the sun of the Stalin Constitution.' "

"Well, in our workers' suburbs," said Julek, "there was also quite a bit of poverty."

"Ha!" snorted Jan indignantly. "In our cities, if you gave a beggar bread without butter, he'd not only call you all sorts of dirty names, he might even throw it in your face!"

★ ★ ★

AFTER four days wandering around our barracks, some trucks came to fetch us. Forty to a truck, we set off, squeezed like sardines. The suburbs, like all suburbs we had seen, combined poverty with desolation. The houses were generally dirty, the streets were messy, the trolleys, as in Lwów, had plywood instead of glass in their windows. Nearer midtown we began to see buildings which showed traces of the former glory of Odessa. We passed a little palace with a garden, then another. We stopped at the garrison baths. The building was miserable and old. The floor of wooden boards was spat on and filthy. The whole place was dark, with very few windows. A big hall served as a dressing room. As in Lwów, our belongings were sent to the famous deskamera. Naked, we marched to the next hall and under the showers. The water was hot. Jan, Julek, Ludwik, and I took possession of one shower, washed each other thoroughly, and left. The dressing-room floor was covered with a thick layer of mud. Julek returned to the shower room and filled a basin with water in which we washed our feet. In the corner of the dressing room, behind a wooden partition, a few men sat among heaps of uniforms, underwear, and boots. We went in one after the other to get knee-high boots made of imitation leather, some long strips of flannel instead of socks, a set of white, numbered underwear, navy-blue cloth pants, a rubakha, a belt, and a cap, the so-called budionnovka (named after Marshal Budienny), made of gray flannel with ear muffs, a star on the front, and a pointed crown. We began to dress. The wrapping of our feet took so long that we were among the last to leave. The sunlight blinded us when we walked out into the courtyard. At the foot of the stairs we, the freshly created krasnoarmieyetz, were received by our comrades with a general outburst of applause. The uniforms so changed people that some of us were unrecognizable. The one who looked best was Rot, an enormous ex-butcher from Lwów. He was literally bursting out of his pants and coat. The budionnovka just managed to balance on his head, which was as big and round as a watermelon. But the fatty was not particularly worried.

"I'll still be having my uniform fitted," he said calmly, "while you're going through the torture of drill on the square!"

The door opened and out came Jan. We almost died laughing. With his slim figure and unusually dignified movements, best fitted for a tuxedo or some costume of the Renaissance, Jan looked incredible in a uniform of the Red Army. He knew it and was furious.

"What are you stupid asses laughing at?" he roared. "D'you think you look any better?"

Jasiu actually gave the impression of a soldier. His uniform fitted well; with his cap cocked on one side he looked quite smart.

When the military employee of the baths appeared, he gave my long "civilian" hair a suspicious look. Taking my cap off, he paled at seeing a full head of curls. If I didn't go to the barber immediately, he said, he would order me to wash all over again. There was no sense in being stubborn. Anyway, I had performed a miracle by having kept my hair for such a long time.

With a diabolic smile, the barber asked me if I'd like my hair cut with a parting, whereupon he let his mechanical razor run down the middle of my head. I shut my eyes; the haircut lasted no more than two minutes. I wiped my bald head with a handkerchief, put on my cap, and walked out.

Before loading, we inspected the vicinity of the baths building. There was a store with vegetables, grapes and apples in large quantities, a grocer's, and a third store with wine and ice cream. This ice cream we devoured all winter, even in a temperature of twenty degrees below zero; but when spring came it disappeared as if enchanted.

The trucks arrived and we began to load, carrying the bags containing our civilian clothes. We were still looking at each other with incredulity. Jan held his *budionovka* in his hand and seriously eyed the red enameled star with the metal sickle and hammer.

"If in September '39 I had been told," he said, "that one day I would wear the Bolshevik star on my head, I'd have placed that man in an insane asylum. What have we come to!" he sighed.

"Jan, keep quiet," joked Julek. "You'd better not start moaning over every star on your uniform; you won't have the time!"

Indeed, the star with the sickle and the hammer was even on the buttons of the slit in our trousers.

Back at camp, we formed a few lines and handed in our bags for storage. Then we returned to our barracks. In the evening they brought us overcoats. While all the clothes we had received in the baths were new, these coats were old and often full of holes. The noncom assured us that these garments were only temporary, as there was a shortage. I wore my coat for a year and a half. Later in the evening the major came to the barracks and announced that tomorrow we would begin to study drill, marching, saluting, and so on, and once through with these exercises, we would be assigned to a battery. Then, ordering the captains of soccer teams to organize a match, he left.

"A kind chap," remarked Lolek, "but a football seems to be the only thing in his head."

The following day after breakfast, fifteen noncoms arrived. Dividing us into groups of twenty, they led us out onto the square and started to give us instruction. It began with parade marching. "Left foot, right arm; right foot, left arm." This was taught us by a *komandir* with two triangles on his collar and a horribly stupid face above it. He demonstrated the parade walk for half an hour, thrusting forth his breast and his behind and beating his boots on the concrete. Finally he placed us in a circle and announced that we would now "try," one after the other. It so happened that he first picked on Rot, the enormous clown, who had long ago decided that he would let them teach him things for a long, long time,

"As long as they are giving me instruction," he reasoned, "they cannot demand anything of me."

To keep from laughing, Rot took a deep breath of air in his lungs and began. Making it very difficult for himself, he walked stiffly, putting down his left foot and accompanying it by a movement of his left arm. The instructor smiled with tolerance.

"No, not that way. Watch it once more," he said softly, politely, and gave another demonstration: left foot, right arm, etc. Gazing at him dumbly with his big, dreamy eyes, Rot marched like a maniac: left foot, left arm; right foot, right arm. The game lasted half an hour, by which time the instructor—having yelled himself hoarse—was nearly out of his mind. But even then Rot calmly continued to walk around the circle in his incorrect manner, repeating softly to the comrades he was passing:

"I could go on walking like this till tomorrow morning. By then I could probably not walk any other way!"

We were all laughing. Finally the noncom ordered Rot back to his place and called on Jasiu. Jasiu marched a few yards correctly. The instructor let out a sigh of relief and smiled with approval. But at that moment Jasiu chose to walk clumsily: left foot, left arm. The instructor roared with anger, jumped up to Jasiu, and began feverishly explaining that he should walk the way he had started. Jasiu nodded his head and repeated the same trick. The unfortunate *komandir* clutched his head in despair. Then Rot struck the decisive blow.

"*Tovarishch komandir*," he cried, "I know how to walk now!" The *komandir's* silly face brightened in a smile of hope.

"All right," he yelled, "begin!"

Rot started out correctly, made a whole round of the circle. Just as we thought he had given up the farce, he started his old game all over again. Throwing out the same arms and legs at once, he walked like an automaton, his face smiling, his eyes filled with tears of concealed laughter. The *komandir* roared like a wounded tiger and ordered:

"Enough! Smoking permitted!"

Thus ended our first lesson in drill, the first of many, for although we had our uniforms, we were stuck in those barracks ten long days, drilling, marching, saluting, and so on.

At last, one cold, cloudy morning at the end of October, we were told that the day of our incorporation into a regiment had arrived. Shortly after dawn we were ordered to form one long rank in the courtyard outside the main block. At our feet lay our knapsacks, and beyond them a few tables stacked with our papers. The listing began. First came questions about education: those with secondary education (almost one hundred of us) were to form one group. Those who had graduated or begun their studies at the university (about forty) made up a second group, They separated us again, then split us into year groups. The divisions were numerous. The segregation they carefully avoided was that based on nationality. After some hours, having been shifted around at least ten times, we were completely confused. We could not understand, for instance, divisions made according to knowledge of certain foreign languages, such as

German and French. Only much later, in conversation with the chief of staff, Captain Lermontov, did we discover that this was for the purpose of statistics, about which the Russians are crazy.

At a given moment a senior lieutenant with a most repulsive face stood up and began jotting down the names of soldiers in groups of professions: plumbers, carpenters, tailors. Then he asked if there were any drivers among us—not amateurs, but professional drivers with actual driving licenses. Russian, of course. I glanced at Jan.

"What do you think?" I whispered. "Should I volunteer?"

Jan nodded his head. "I guess so," he said with reflection. "Better to drive a truck than clean guns."

I was the only one to raise my hand. The *komandir* came up to me.

"Do you have your license?"

"I do," I replied, and handed him the document.

I was a second-class driver, which in Russia signified that I was also an experienced mechanic. The lieutenant looked at me with contentment, told me to take my knapsack and stand aside. I had a feeling that I had made a mistake, but I was not sure how serious. After a while a battery chief read out a few names from a list—plumbers, carpenters, tailors, and myself. He then led us to the central block. On the door of a small barrack room, I read a notice: *H.Q. Battery*. I was furious at myself; I could have anticipated such an outcome. I was given a bed, with a straw mattress already made. The beds were double-deckers. On the deck above mine I recognized Misku, a boy I had known in Lwów. One day at the beginning of the winter of 1939 he had repaired a tile stove in the room I occupied with my father.

"What?" I asked. "Are they looking for tile workers too?"

"Oh, no," smiled Misku maliciously. "I said I was a carpenter. What kind of work could there be for a carpenter in such a mechanized regiment?"

I knew then that by saying I was a driver I had made a great mistake. How could I do such a stupid thing as to confess my driving capacity without first of all making sure what would happen to the rest of us?

★ ★ ★

WHILE wondering how I could correct that mistake and get transferred to my comrades, I made use of my official duties to acquaint myself with Odessa. It is a beautiful city. The architecture along the seashore—formerly the aristocratic quarter—is very pleasing to the eye. On the Feldman Boulevard I found the Hotel Inturist, designed for foreigners and Russian big shots. I don't think there were any foreigners in Russia at that time, but the Russian dignitaries in this hotel were living in a state of luxury not one Russian in ten thousand had ever dreamed of. One day it was my job to take some cases from the station to the Inturist. A waiter who spoke French showed me the menu. It was no more modest than a menu in the most elegant hotels in the "rotten capitalistic world." Two blocks away, leading to a store, I watched a mile-long line of men and women waiting—with ration cards, of course—to buy potatoes and bread. The citizens of a country which at one time used to be the granary of Europe received four hundred grams of black, clayish, badly baked bread. This was far too little for people who consider bread their first and most basic food. I am sure that Americans do not eat more than two hundred grams of bread a day, but in America bread is merely an addition to a meal. At home I did not eat more than two or three rolls and perhaps a slice of bread a day; but in Russia, even in the army, my portion of four hundred grams was not sufficient for me. In the same store, sweet white rolls were sometimes obtainable, but an ordinary citizen could not buy them; at forty kopecks each, they were too expensive. A worker earned between three hundred and five hundred rubles a month, depending on his qualifications. With this he had to feed his usually numerous family. For in Russia, as in most totalitarian countries, mass procreation of children is a patriotic duty, recompensed by the state. Any interruptions of pregnancy, or abortions, were severely punished. When you consider that a man's suit costs him a thousand to fifteen hundred rubles, his shoes two hundred rubles, to buy which he had to wait his turn on a long list for an allotment, you can imagine how well he lived and fed his family on three hundred to five hundred rubles a month.

Some time later in Odessa I met a university professor, a well-known biologist. His monthly salary amounted to eight hundred rubles. His only privilege was that he lived in one-half of his old pre-Revolution apartment, the other half having been given over to some statistical bureau. This house was located in the lovely seashore neighborhood of Arkadia, where the vacation homes for workers had been built. These homes, surrounded by amusement parks in whose high grass and shrubs free love proudly flowered, constituted the one great and visible improvement the regime had made. In the springtime, Arkadia, covered with superb, juicy, Black Sea verdure, seemed like a land out of a fable—provided you avoided its main throughfare, in which hundreds of radios, gramophones, microphones, bands, amateur singers, and accordion players competed with one another at the game of shattering nature's silence.

While gazing at all this, my predominant sensation was one of sympathy. Never a model altruist, I could not help feeling sorry for these people. My former reluctance, even a certain dislike stemming from my experiences in September 1939, and the difficulties caused by the Russians in our country during the past year, changed now to sympathy.

But I had more important things to do than to consider the Russian problem. There was my own, that of my transfer. I had discovered that nearly all our boys with a secondary education had been grouped together to form a brand new battery in the Third Division. The battery carried the number nine. On my first visit to my comrades, I found an almost night-club atmosphere in their dormitory. They were all delighted with the outcome of their assignment. All the battery's enlisted men were from Lwów. The few Russians were apparently rather well selected. The younger officers, even the political leader (who accompanied every battery) gave a good general impression. And the *kombatr* [abbreviation for battery commander] Kalugin, was "supposedly a European," as Jan put it. Anyway, he knew French and German and held Western culture in great esteem.

When I learned all this, I felt sad and mad. Jan tried to console me. So did Ludwik, who kept repeating that I simply had to get myself transferred to their battery.

"After all," he kept saying, "the only condition is secondary education, and that you've got, since you've studied at the university."

I agreed with them, of course, and from then on every evening I went to the regimental office to inquire about this transfer.

On the eighth day they admitted me to Captain Lermontov. At first the captain asked me whether I was aware of the fact that my behavior was irregular and against regulations, and that in any case, regardless of the result of the conversation, I would be sent to the guardhouse. I agreed to the guardhouse if he would consider my petition. He laughed kindheartedly and asked why I insisted so much on being transferred to the Ninth Battery. Looking straight into his eyes, I lied shamelessly that it was my dream to become a reserve officer of the Red Army. I emphasized the word "reserve." Lermontov said that the matter of transforming the battery's status into an officers' school had not yet been confirmed, but that he could understand why I was longing for the Ninth Battery.

"Of course, you miss your friends." He smiled.

I nodded.

"Well, but we are short of drivers," hesitated Lermontov.

I declared that if necessary I could drive for the whole regiment, while still being assigned to the Ninth Battery. The captain thought for a while and agreed; he told the secretary to write my transfer in the order of the day. In the army it is forbidden to say thank you for anything, so I just stood there with a joyful face and waited for him to order me out of the room. Unexpectedly Lermontov stood up, took his cap from the hanger, tapped my shoulder and said:

"Come, you will take me home."

I put on my *budionnovka* and, utterly surprised, followed him to the corridor.

"I'm now going home for half an hour," said he. "If you want a good cup of tea, come along."

Naturally I accepted the invitation. We left by the guard's booth. At the fence, Jan, Ludwik, and Walter were devouring some grapes I had brought them from the city and sadly spitting out the stones onto the street. When they saw me walking with Captain Lermontov, they opened their eyes wide. I waved my hand to them with condescending friendliness and marched down the street with Lermontov.

The captain lived in a small house on the corner of Kiev Street. He had a four-room, rather decently furnished apartment. His furniture did not look Russian. He noticed that I was inspecting a desk made of steel tubes.

"From Finland." He smiled with contentment.

I found out that almost everything in the house had come from Finland, For his part in the Finnish campaign Lermontov had been decorated with the Red Star medal; he was also a member of the party. He was only twenty-six years old.

"Irina, come, we have guests," he cried in the direction of a closed door. The door opened and a rather short but beautifully built girl appeared. She had blond hair, cut short, long green eyes, and a dark, soft skin. She looked at me without embarrassment, rather with insolence.

"My sister," presented the captain. Then he turned to the girl.

"You wanted to see a *zapadnik* [Pole ("zapad"-west; "zapadnik"—person from the west)]. This one will go to the guardhouse tomorrow. Give us some tea."

While the girl was in the kitchen preparing the samovar, Lermontov showed me an album of photographs of the Finnish front: the reinforcement on the Mannerhein Line, Vyborg ruined by bombing, the Paetsamo Lake, the regimental guns poised to attack near Vyborg. Then Irina came in with a samovar on a silver platter.

"Tea will be ready immediately," she said.

She had a soft, low, sensuous voice. I did not expect such a voice from her. I must have made a stupid face; she looked at me and asked in a low, melodious tone:

"Are all of you so shy?"

I looked insolently into her eyes, I said that one could easily become dumb from admiration, looking at her, and added that I was the most timid of us all. Lermontov was amused.

"Don't you play up to my sister," he said to me; and to her, "He is so shy that for eight days they have been kicking him out of the office, but he keeps coming back, every time under a different pretext."

The girl was pouring tea into thin china cups, which surprised me, as Russians usually drink tea from glasses. I examined the cups carefully.

"Also Finnish," explained the captain with a shameless smile.

Well, I thought to myself, they evidently did even more robbing in Finland than in our country.

Lermontov must have guessed my thoughts. "We did not have to buy a thing," he said.

Irina served raspberry jam on little plates. We drank tea in silence, interrupted only by the lip-smacking of Lermontov, who swallowed from the saucer.

Feeling the girl's curious look on me, I raised my head toward her. She was not at all disturbed, but continued looking at me insolently, with a vague smile. I was the first to take my eyes away.

The conversation turned on the subject of Lwów, Cracow, and life in Poland before the war. I talked with great reserve, determined not to create the impression that I wanted to underline the great differences between then and now. The phone rang in the hall. Lermontov exchanged a few words with someone, came back to the room, and said to me:

"I have been called to the major. If you wish, you can stay a little while longer. Irina," he said, turning to the girl, "take him to the guard's booth later, or they won't let him in."

I mumbled some words of thanks and good-by. The captain grabbed his cap and belt and hurriedly left. All this time Irina had been observing me with that enervating smile around her lips.

"You give me the impression of being afraid to stay with me," she said in a soft voice.

I looked at her. The catty green eyes, half veiled by long lashes, were mockingly smiling. Funny girl, I thought.

"If one of us is afraid of being here together, it is surely not I," said I.

"I'm not afraid. You wouldn't murder me, would you?" said she. The green eyes went on smiling. Damn it, what was she up to? Being provocative, or simply flirting? I couldn't understand her behavior. I was reminded of a warning my "master" in the driving profession, an old Lwówian taxi driver, had once given me: "Keep away from the wives, daughters, and fiancees of bosses." I retreated a little with my chair and asked vaguely about social life in Odessa. She looked at me coldly and began talking in an indifferent tone. Actually, there was no social life: she went to the opera, to the theater, the movies; from time to time to a dancing party at her college. What did she study?—Painting. I began telling her trifles about life in the West, coloring my stories as much as my vocabulary permitted, helping myself out with French, of which she knew a little. I showed her some photographs. She looked at everything with curiosity, and drew up so close that a strand of her hair touched my cheek. A seducer, I decided. I put the snapshots in my wallet and got up from the chair. She lifted her head and looked at me inquiringly.

"I have to go," I declared. "It's nine o'clock."

Irina stood up without a word and brought her coat from the hall. I took it out of her hands and helped her put it on. While doing so, I happened to touch her naked neck. She stood stock-still, as though paralyzed. Then she turned her head toward me, her eyes half closed. Nymphomaniac, I thought, changing my opinion.

"I am sorry to trouble you," said I politely.

She shook her head.

"It is an honor for me," she said almost ironically.

We went out into the street. I didn't feel like talking. I was furious. Irina also seemed to be mad at me. After some fifty meters she took my arm and smiled coquettishly.

"Come to our house from time to time; we can play cards or chess," she said in a begging tone. "I am so bored at home alone."

Of course I promised that whenever I could I'd come.

"Just call me and I'll tell my brother to give you a pass. We can even go to the movies in the city," she assured me.

I thanked her. Now that was a different question. To secure a pass to the city with her pull was not to be disregarded. We reached the gate to the barracks. A noncom came out of the booth and saluted Irina. She was about to explain my return when he forestalled her:

"The captain has already warned me that a *zapadnik* would turn up and that I should let him in."

She shook my hand.

"Don't forget your promise," she said.

I entered the booth; the soldier was laughing loudly.

"Aha! Lermontov's sister. You'd better be careful, brother. She's fond of the boys. More than one here has had troubles because of her."

"What d'you mean by troubles?" I inquired.

"Well," explained the soldier, "they got stupid, came back late from their leaves. Not even the captain could help in such cases."

He then proceeded to tell me what had happened two months before our arrival. In the H.Q. Battery (mine) there had been a journalist from Moscow—a handsome, intelligent, and well-informed young man with a very good reputation among his commanders. He was to have finished his service on September 1 and returned to Moscow. In the summer, however, he began to frequent Irina Lermontov. The girl wrapped him up so completely that he ceased to be interested in his work, neglected his duties. Finally, at the beginning of August, he received a pass (from Lermontov, of course) for three days. He stayed in the city, with Irina, naturally. On the day he was due back in camp the couple had quarreled. He returned only the following morning, completely drunk, though he had never been a drinker. Within twenty-four hours a court martial had sentenced him to a year's *shtrafbatalion* [penal battalion, whose members, receiving uniforms and food of the worst kind, had to do eight hours hard labor and five hours normal military drill per day].

"If he's strong enough to live through the penal battalion," concluded the soldier, "he'll be back again in a year. Then, before he can return to civilian life, he'll have to make up for the months of service he missed. During the examination he maintained that she had set up traps for him. Strange girl, that Lermontov sister!"

I shrugged. This was none of my business. Nobody had ever set up traps for me.

★ ★ ★

NEXT day, amidst triumphant shrieks from my gang, I dragged myself, my knapsack, and my straw mattress up the stairs of the Ninth Battery block. There I placed my things on the floor and went to the office to report. Lieutenant Kalugin was sitting at his table. I clicked my heels and began to recite the old story: "Tovarisch lieutenant, I report that Krasnoarmieyetz Virski has been assigned to the Ninth Battery." At the next table I saw Ludwik, who had managed to become secretary to the sergeant major, Shaposhkin, a man only vaguely acquainted with the complicated art of writing. Ludwik lifted his happy face from above some papers and turned to Kalugin.

"This is our famous automobilist," he praised me seriously. I found out that my comrades had been trying on their own to bring me back to them by working on the *kombatr*, telling him fantastic tales about my having raced in automobile contests. Kalugin was visibly pleased; he shook my hand, asked me what types of cars I had driven. In reply, I cited in one breath the names of fifteen brands of European and American automobiles, none of which I had ever driven in a race. Delighted, Kalugin asked me what platoon my friends were in. Ludwik answered for me.

"Sergeant!" said Kalugin to Shaposhkin. "Give him a bed among his friends!" And to me, "You will be the driver of the battery's headquarters; I will ride with you."

I saluted and marched to the dormitory. They gave me a bed on the upper deck. Under me were Jan's and Julek's; my neighbor on top was Misha, a very nice boy, a student of architecture, who spent all his spare time examining 150 pictures of his first love, a Russian woman ten years his senior.

The first evening with my friends was celebrated with a bottle of vodka, the last of the Lwów supplies. I went to bed with a feeling of comfort, knowing that I was again among friends.

Suddenly the guard roared: "*Odboy!*"—which meant not only "Lights Out," but complete silence. The chattering was cut short. Someone from a bunk further away let out a long, sad rectal sound. The immediate reaction was general gaiety. Shaposhkin stuck his head into the dormitory.

"Did you hear *odboy* or not?" he whispered.

The giggles stopped, but the same sound was heard a second time, then a third, a tenth, until they became too numerous to count. In complete silence (I mean, no one said anything from his mouth) the entire battery competed in the manufacture of these sounds. In the Red Army regulations there were no specifications against producing rectal explosions after *odboy*. Under me Julek was performing such miracles that my own blankets were waving. Shaposhkin stood in the doorway with a startled expression on his face, but he did not say anything. The following evening after *odboy* the same thing was repeated. Finally the *Natzmen* [Russian slang for Soviet citizens of Asiatic origin] H.Q. Platoon, which slept opposite us, grew very indignant over this lack of manners, and complained to Kalugin, who asked our battery what it was all about. Ludwik explained that our stomachs, unaccustomed to black bread and borscht, reacted in this way and that it was not our fault. Kalugin smiled. That night after *odboy* the cannonade grew even stronger, but this time the game took a different turn. As soon as we had finished our evening performance, and, with a feeling of a duty well done, were trying to go to sleep, we heard the revenge from the H.Q. Platoon. It is hard to describe that din. I don't know whether the letting off of intestinal gases is a national custom of Uzbeks, Tadziks, or Turkmens and that this is why they are such experts in that field, but they paid us back in such a way that the next evening none of us tried to start a fight for pre-eminence. We considered ourselves defeated and signed with the *Natzmen* a treaty of eternal peace.

Life in the battery of "brothers," as we were christened by the rest of the regiment, was becoming quite bearable. The noncoms were really very decent and our relationship with them more than correct. The *komandir* of the reconnaissance detachment, to which I belonged, was Dunin, a rather intelligent, pleasant man from Kiev. My boss at the automobile lot was Soloviey, a peasant idiot, but not dangerous, least of all to me, as I had very little to do with him.

My occupations from now on were radically changed. I was assigned a one-and-a-half ton truck, which was supposed to sparkle with cleanliness and to be used only on maneuvers and, naturally, in case of war. Standing merrily on wooden blocks, it looked like a truck out of a shop window. For a time Soloviey inspected the condition and upkeep of the material, but in such unexaggerated a fashion that we could not complain.

Two drivers were assigned to the battery: Shevtchenko, an effeminate youth with an infant's face, who came from a "kolkhoz" (collective farm) not far from Odessa; and Vania, his inseparable comrade, also from a kolkhoz somewhere in the Ukraine. Vania was well-built and handsome, but he had a vacuum in his head and considered Shevtchenko the oracle on all questions of technical detail. Among our boys the candidates for drivers were Lolek and little Zyga, who always followed Lolek and me around like a dog, unable to take one step without our advice.

The trucks of the entire division were parked together. Next mine was one from the H.Q. Platoon. Its driver was an Armenian, Aram Babayan. Broad-shouldered, short, with a tanned face, wide mouth, high cheekbones, and a big, curved, oriental nose, he had, nevertheless, a pleasant appearance. Aram had worked as a city hall employee in a small town near the capital of his Armenian republic, Yeryvan. Though a graduate from secondary school, his education registered zero. He could manage the mechanical work at the city hall, he could drive, and he knew something about auto mechanics, but otherwise he was a complete ignoramus. Aram was a primitive man. My friendship with him began in a strange way. At the beginning, working side by side, we rarely communicated; sometimes we would offer each other a cigarette or borrow tools. One morning Aram and some other drivers from the 2nd Division got very drunk on a bottle of *sivukha* [low-grade vodka]. At noon a messenger appeared with an order that all trucks of our Third Division were to be driven up to the main workshop for a technical inspection. The inspection was to be taken by Captain Kovalski himself. The division's military technician, Piltozov, and his aides from the various batteries were also to be present.

Aram was unconscious, lying on top of a truck. In that state, of course, he could neither drive nor present himself before any of his superiors, since that would cost him at least ten days in the guardhouse. From among our men, the first to drive out to the inspection was Shevtchenko; my turn was to come after Vania. Shevtchenko came back saying that the commission paid no attention to the driver; they just took the number of the truck. I told Lolek and Zyga to take as much time as possible after my departure and to let the drivers of other batteries go first. My inspection lasted only a few minutes; I returned, and after letting two or three other drivers go, I went a second time in Aram's car. Kovalski recognized me immediately and smiled.

"What's this?" he asked in Polish. "Are you driving instead of others?"

I replied that one of my comrades, candidate for a license, did not have enough courage to drive himself, so I replaced him.

"Where is he?" asked Kovalski unexpectedly.

I quickly lied that, not knowing how to drive, he was ashamed to get in the truck with me.

"Oh, how bashful he must be!" laughed Kovalski, "like a virgin before her fifth child!"

I came back to the parking place. The Armenian was still on top of the truck, with terrible eyes and unconscious.

We went to dinner. Later that evening Aram came up to me in the hall. Without a word he put out his hand, held mine for a while, and said:

"Thank you. It was because of you that I avoided punishment. None of my old comrades did this, while you, practically a stranger, helped me. I am now your friend and servant." He spoke in a broken Russian typical of the peoples of the East. Before I had a chance to answer he had disappeared.

From that day, he showed me his friendship in a way often incomprehensible to me, but always careful, almost tender. Whenever he managed to steal something from the kitchen he would share it with me in a brotherly fashion. When I sometimes came to the parking lot later than he, because I had drill exercises that he did not have, I would find him cleaning my truck. Sometimes, seeing a hook or a button missing on my coat, he would get up before the bugle, which was severely forbidden, and sew it on. When I fought with one of the drivers who had gone off with one of my tools without asking, I would find Aram standing behind me in silence. If the quarrel dragged on, he'd say shortly, gazing menacingly at my opponent:

"You'd better shut up or I will make you."

His terrific strength was notorious in the regiment and no one wanted to pick a bone with him; consequently in his shadow I felt safe and sure of myself.

One day Kalugin cursed me violently in the corridor for coming into the barracks in muddy shoes. We were supposed to wash our shoes in the courtyard puddles. Aram, standing by me with a furious expression on his face, suddenly snapped:

"He is not guilty, *tovarisch kombatr*, and there's nothing to get excited about."

Before Kalugin could recover from this daring insolence, Aram had disappeared down the corridor. The *kombatr* gazed after him in surprise, then turned to me:

"Is he your friend?"

I nodded. Kalugin could not get over his admiration.

"By what a miracle did you earn the friendship of an Armenian?" he kept asking me, altogether forgetting his recent anger.

From Kalugin I found out that Armenians, like most peoples from the East, are not eager to establish friendships. But when they do make friends, then you can count on them in every situation. It is a trifle for an Armenian to stick a knife in a man's back, and to do so in the defense of a friend is an honor.

What was happening to the other boys? There were 230 of us, 96 of whom were in the Ninth Battery. The rest were dispersed among all the other batteries of the regiment. Our battery was the center of news from Lwów, as well as a sports center. Misku, my companion from the H.Q. Battery, came almost every evening to chat in our dormitory. "The mocker's loge," as we called our few beds, always attracted a group of listeners. Misha read his poems, which he himself called futuristic, and which we called hopeless and super-filthy. Ludwik, with his typical morbid humor, described scenes he had witnessed during the day in the office; Jasiu told fifty-year-old jokes, for which Walter would hit him on the head from the upper bunk. Aram always squatted by our beds and, without understanding a word, laughed loudly when he saw that we were laughing. When I tried to translate a joke he would interrupt me, saying that it was not important, that he was happy so long as he saw us happy.

Apart from myself, none of the men in the battery had gone to town, except to the baths, where we all went once a week. Until we took the pledge we were not allowed out even with a pass. During this period, so it seemed, we were not under the jurisdiction of military law.

Irina sent me two notes by messenger, asking why I had not come. I answered neither. Finally, one day after dinner as I was returning to the barracks in overalls, I ran into Lermontov standing with a few officers and Irina. As I saluted, Lermontov smiled and attracted Irina's attention to my presence. She turned around and merrily waved her hand. I saluted again and strode on, making straight for the open barracks door.

"Fred!" she called.

I was surprised to hear that she knew my name; I did not remember having told her. I stopped and gazed at her inquiringly. She came up to me.

"What? You don't even want to come and say hello?" she asked, passing, I do not know on what basis, from the formal "you" to the familiar "thou."

I explained that I was in a hurry to get to the barracks to wash and change. She disregarded that.

"Put your uniform on and come down. I'll wait. You shall have supper in our house. My brother is inviting you," she added quickly, seeing that I had a face indicating that I was going to refuse. Lermontov shouted from a distance:

"Hurry up! We will wait for you here."

There was no use refusing. I ran upstairs, washed fast, put on my uniform, and reported to Shaposhkin that I was going to supper at Captain Lermontov's. The poor man opened his mouth wide in astonishment. I went downstairs. The group was still waiting. As I came up to Irina, I heard a terrific screaming from the second floor. The entire battery was in the windows, everyone howling himself hoarse. Irina did not seem embarrassed; she lifted her head and waved her hand. Lermontov looked at the smiling faces in the windows and also smiled. By an odd coincidence, in the guard's booth I met the same noncom who had once warned me against Irina. Now as I passed him, he whispered in my ear:

"Don't forget what I told you!"

Supper at Lermontov's was spent in a strange atmosphere. Considering local conditions it was luxurious: apart from caviar, which in Odessa was no more expensive than bread, there was meat, cookies, and the inevitable tea. After supper Lermontov and the other officers said good-by and went out to play cards with friends. I was left alone with Irina. She behaved as she had the first evening, provocatively and insolently. Around eight I grew bored, wanted to say good night and leave. She laughed artificially, but not without satisfaction.

"You cannot return to camp by yourself," she said, "and I cannot go out until my brother returns. That will not be before eleven."

I was imprisoned. She again began her old game. I was coldly polite, though furious. After all, I was only twenty-one and not made of stone. Finally, she began to deplore the men of the "rotten" West. She expressed her suspicion that I must be either homosexual or impotent. That was too much. How could I allow her to have such an opinion of me?

★ ★ ★

WE began preparing for the October Revolution parade almost a week in advance. Because of the difference between the old and new Russian calendars, this holiday falls not in October, but on November 7. For hours every day the battery marched up and down the yard, shouting "Urra-a-a-a!" in front of a provisional tribune. The morning before the parade we received steel helmets, new coats, and boot brush and polish for the occasion. Before we actually passed the tribune we must have shined our boots at least a dozen times.

It was still night when we entered the square. Only at dawn did we get some idea of its size and the vast number of troops it held. Julek estimated the mob at eighty to a hundred thousand men. Until eight o'clock we just stood there, cursing the Revolution, the army, and all parades. At eight sharp the command "Attention!" was bawled over the loudspeakers, followed by some short bugle calls, one lone voice, and then a horrible roar: "Urrrraaaa!" This was repeated once every minute. Then General Tcherevitchenko, the commander-in-chief of the military district of Odessa, came galloping through the tightly knitted regiments on a huge white horse. Rearing the animal in front of each regiment's banner, he saluted and roared:

"Long live the holiday of the October Revolution!"

To which the regiment of a thousand men took a deep breath and yelled "Urrrraaaa!"

Although we had already howled our "Urrraaa!" and the general had galloped away, we had to remain motionless until the last "Urrraaa!" had died away and the loudspeaker given us the command: "At rest."

The inspection and the shouting over, a military band struck up and the parade itself got under way. Slowly, one after the other, the regiments began marching out. When our turn came we moved in close formation to our starting line until we saw a free space of concrete and a tall tribune on our right. As high as a two-story building, this tribune was covered with red cloth and a tremendous emblem. In the center, surrounding the hammer and sickle on the globe, hung a great wreath made of ears of wheat. On the tribune stood thirty or forty soldiers and civilians, and at its base the guards, with automatic pistols slung across their breasts. The concrete resounded to the slow, thunderous tempo of iron-heeled boots. The battery before us had just finished its yelling when, passing the tribune, Kalugin lifted his hand to salute. On that signal our first rank began its "Urrrraaa!" which swept backward in a wave. I would never have believed that a hundred throats could make such a din. The noise resembled that of a sea billow, particularly in its rising and dying out. Kalugin, walking backward, stared with unconcealed pleasure at the battery as it marched by.

We reached camp at two in the afternoon—very hungry because we'd had nothing to eat since breakfast at three-thirty that morning. The dinner was exceptionally rich, festal. At four the next ceremony began—the pledge. This took place in the courtyard, where they had placed a big gun, in fighting position, decorated with flags and the regimental banner. Everyone had in his hand the "Red Army Man's Book," in which the text of the pledge was printed. "I, soldier of the Red Army, swear before the nation, etc." It was about loyal service, readiness to fight the enemy to the last drop of blood, refusing to surrender, keeping military secrets, and a whole lot of other obligations.

"Well," said Julek, shaking his head, "if I keep all the promises written out here, you can call me Maximilian."

The major and commissar came to the center of the block, where the major made a long speech on the importance of a soldier's pledge. He emphasized how proud we should feel as the first citizens of "former Poland" to have the honor of being soldiers in the Red Army.

"May you choke with that honor," I heard Jasiu mumble behind me.

Then the commissar spoke. Having repeated something along the same line, he emphasized all the punishments that threatened those who dared break the pledge. For any transgression of the code you could get several years in a fortress, and during the war, of course, you'd be shot. We then took off our caps, raised our right hands, and repeated in a chorus the words of the pledge. For ten minutes, while we presented arms and the officers saluted, the regimental band played the "International." Opposite me stood Jan. I had noticed at the beginning of the anthem that he had been grimacing and making frantic gestures. When the celebration was over and we were moving toward the barracks, he whisked off his stiff "holiday" cap. Though strictly against regulations, he walked with his cap in one hand, rubbing his shaved head with the other. I looked at his skull. On the very top I saw a big red bump. Jan was swearing violently. It appeared that after the pledge, when he put his cap back on his head, a horsefly or a bumblebee had gotten under it and, during the entire "Internationale," terrified by this sudden imprisonment, had been stinging his scalp.

★ ★ ★

THE days passed monotonously, days filled with camp life: drilling, polishing trucks, and attending the politzanyatya—our so-called political education.

For this informative and "educating" work, five to eight hours of the weekly program were sacrificed, these "lessons" were conducted by the *politruk* (political leader) or his aide. The topic was usually a story of the victorious battles of the Red Army after the Revolution—the battles on Lake Hasan somewhere in Mongolia, the war in Finland, even the campaign in Poland in 1939. Another topic was the history of the Russian Bolshevik party, very interesting, but frightening in its bloodiness. It was amusing to hear the *politruk* explain to us such concepts as democracy, capitalism, bourgeoisie, etc. All his explanations had been studied by heart according to an unchanging formula. From one lesson to the other he assigned to us paragraphs from the Stalin Constitution, which on paper contained beautiful—but in practice nonexisting—privileges, such as freedom of speech, press, and religion!

Another of the *politruk's* duties was the censorship of all incoming and outgoing mail. This was done rather discreetly, without any notations on the letters or crossing out of words or sentences.

Most of the time a letter that contained something "impious" did not reach the hands of the addressee. Sometimes the receiver was held responsible and had a hearing. Three months after I was inducted into the army I received a card from Lisbon from a young girl I had known in Cracow. The card had been addressed to Lwów and forwarded to me in the regiment. It was adorned with a mass of different stamps and annotations. Among the stamps I found some from Paris, Berlin, Warsaw, Moscow, and others. There was even a paper extension added to the card for all the stamps that could not fit on. The card contained a few items with little meaning: greetings, a question about the girl's father, who had remained in Poland, and so on. That same night at around one o'clock, a soldier on duty woke me up, telling me I was wanted by Shurin, the regimental commissar. I dressed quickly, having not the slightest idea what Shurin wanted. He smiled at me and pointed to an armchair opposite his desk. He handed me a box of cigarettes and offered me coffee. I was amazed at these Oriental ceremonies. When we had lighted our cigarettes, the commissar asked me how I was getting along, how I liked Odessa. I answered with much reserve, waiting for what was to come.

"Who wrote you this card from Lisbon?" Shurin suddenly asked. At last I knew what he was up to! I should have known right away! Mail from abroad! From the "rotten bourgeois West." I concentrated my attention, even though I had nothing to hide. Well, the card is from a friend, a girl from my home city; she is now sixteen years old.

"Why is she in Lisbon?" asked the commissar suspiciously.

I shrugged.

"How can I say? All I know is that she ran away from the Germans."

"And what does she write?"

I repeated from memory the contents of the few lines. Without embarrassment, the commissar took out of the drawer photostatic copies of both sides of the card.

"Translate literally," he commanded.

I translated word after word, though I was certain Shurin understood the original. His secretary jotted down all my answers. The commissar asked me another few questions, apologized for having disturbed me in my sleep, and told me to go. I didn't know how long I had slept when another messenger woke me up. I was wanted by the commissar. Cursing violently, I dragged myself to the central building. It was 4:30. The scene of four hours ago was repeated in precisely the same form. Again the kind smile of Shurin, the sleepy secretary in the corner, cigarettes, coffee, the same beginning of the conversation, same insignificant subjects, same questions concerning the card and its sender. The situation was so deceptive that, in my drowsiness, I thought I was still at the first hearing.

But no, the game ended, the commissar shook my hand and told the messenger to inform the sergeant major that I was to be excused from all occupations that day and that I might sleep as long as I liked.

The next night the same thing happened. The same questions, the same post card on the desk. Again two hearings, one at two and the other at five. Smiles, the sleepy secretary, cigarettes, coffee, and "Why is she in Lisbon?" etc., including the excuse from work for the next day. This absurd and unintelligible procedure lasted three days in a row; then there were three days of rest and again two days of hearing. Probably the commissar figured that eventually I would make a mistake in my replies and confess something. I couldn't do that even if I had wanted to, as I had nothing to conceal.

★ ★ ★

AT the beginning of December we were introduced to the so-called "Voroshylov runs," a performance practiced by all Russian regiments at that time. For these runs the regiment batteries formed teams. You started off by running six kilometers, then nine, then twelve, and finally twenty-four. The pace was known as the "marching run," a combination of walk and run, with full equipment and rifles. The winning battery received prizes in the form of additional passes, and so on. For victory they took into account not only such points as speed, but also the manner in which the battery kept its original formation. And they subtracted points, of course, for every soldier who dropped out. It was a sinister sport. Running, particularly in mud up to the ankles, was never among my predilections. How to get out of it? We worked out a number of ways, but none seemed satisfactory.

The day of the first run came. Starting at the camp entrance, the route turned a right angle onto a street leading to a crossroads, and from there onto a road which led to the steppe. At this crossroads there was a prison, and opposite it a few houses with shops, among them a liquor store with billiard tables. We came to a decision: the liquor store was to become our refuge. Lolek and Julek and I were to take a chance on this very risky enterprise. The batteries started out one after the other in one-minute intervals. At the gate, where the starting and finish lines were marked, stood the judge's staff—the commissar, the major, and others. Starting off to the inevitable "Urraa," the first battery passed through the gate at a canter, sinking deeply into the quaggy mud. Julek, Lolek, and I took our places on the extreme right of the column, one behind the other. At last it was the Ninth Battery's turn. With a shout we set off onto the street. Kalugin was a great sportsman and it was his ambition to see his battery come in first. We therefore ran at a sharp trot for the first five hundred meters. The mud splashed over our uniforms, guns, and faces. After a hundred meters or so we looked horrible. Before the crossroads Kalugin again broke from a walk to a trot, so that we drew close to the Eighth Battery, which had started out before us. At the crossroads there was a tremendous puddle. In spite of our efforts, the formation was upset; everyone had to avoid the puddle, as it was very deep. I turned around. Julek and Lolek were running on the sidewalk with very innocent faces. I followed their example. All the officers and noncoms were up front, so that the rear of the battery could not be seen.

"Hop!" screamed Julek.

Together we leapt from the ranks to the door of the wineshop. In one second we were inside. The door slammed after us. We could hear the tramp, tramp of boots and the "Urraa" shouts in the distance, as a battery started out. In the shop there were a few people; they stared at us, smiling. The salesman asked:

"Well, you don't like running?"

"We do," answered Lolek seriously, "but we get dizzy when we do too much of it."

The salesman laughed. "Well, I hope nobody catches you."

He scratched his head and led us to the rear room with billiard tables.

"Nobody will come in here," he assured us.

We sat down, wiping the mud off our faces.

"Well," concluded Julek, "if we manage to join the battery when they return, we have won."

After half an hour we heard shouts approaching. The First Battery was returning. We peeped carefully through a slit in the door. The men looked terrible, covered with mud from their feet to the pointed crowns of their caps. In a few minutes our battery should be passing. Lolek took us out onto the courtyard, where there was a great puddle. He looked at us with resignation and without a word lay down in the mud, wallowed in it, and got up again. Then he began smearing the mud all over his face. At first we thought he had gone crazy, but soon realized he knew what he was doing.

So we followed his example, smearing ourselves thoroughly with mud. Julek was standing at the door, observing the order in which the batteries were coming back.

"Ours is just coming," he whispered. "When I wave my arm, jump into our rank!"

We were standing behind him and could not see the street. Finally we heard "Urraa" shouts, louder than before. Then Julek waved his arm. The three of us leapt out onto the sidewalk and mixed with the crowd of muddy uniforms. Since everyone was again trying to avoid the tremendous puddle, all formation was upset. Lolek, in front of me, was howling like a fiend, moving his short legs at an amazing speed. Nobody, it seemed, had noticed our sudden reappearance. A few seconds after the first excitement, I looked around. The first face I saw was that of a Russian, but not from our battery. I turned quickly, and raced on, realizing, to my horror, what had happened. We were among the soldiers of the Eighth Battery! In his excitement, Julek had not counted properly and waved his arm too soon! Lolek turned his frightened, muddy face to me.

"We are done for!" he sobbed. "What will happen now?" Julek cried, "Don't worry. We'll say our battery's tempo was too poor for us!"

We were not convinced, but all we could do now was to finish out the run. The soldiers around us were breathing heavily, their faces perspiring. We, of course, were in perfect shape. At the gate stood the judge and jury: the commissar, stop watch in hand, the major counting the rows of four as they passed over the line. When he saw our faces, he was so surprised that he stopped counting. The battery halted and the major shouted:

"What are those three *zapadniks* doing here?"

We dropped out of the formation.

"*Tovarisch* major," reported Julek, pretending to gasp for breath, "our lot were running so slow that we caught up with the Eighth Battery and returned with it."

The major, who had a very high opinion of our sporting abilities, stared at us with appreciation.

"Good sportsmen," he concluded. "You will get a pass today." We saluted and went to the wall. Kovalski, who did not question Julek's or my achievement, was looking critically at Lolek's short legs.

"How on earth did you manage it?" he asked. Lolek modestly lowered his eyes.

"One does what one can," he said sweetly.

Kovalski was sure there had been some irregularity, but he could not see what. Then our battery with the roaring Kalugin at its head arrived on the yard. Having made good time and kept good discipline on the run, they received second place in the regiment. With our "deed" we added quite a few points to the result. Kalugin was enchanted with us; he hugged us. Only after a little while did he begin to think aloud:

"How is it that I did not notice you when you were passing me?" he asked.

Julek quickly explained that it happened at the crossroads, in the general confusion due to the puddle. This explanation seemed sufficient for Kalugin. Deciding to kill two birds with one stone, I said to him:

"*Tovarisch* lieutenant, I will be able to run nine kilometers again, but no more, because on long distances I have heart palpitations."

Kalugin smiled kindly. "You must remind me," said he, "and I will excuse you."

Kalugin was a decent, intelligent fellow, but, like all of us, he had his weakness. For some time now we had formed the strictly illegal habit of staying in our dormitory at suppertime. Instead of standing outside in the freezing cold, waiting to be led into the mess hall, we took it in turns to go and fetch the bread and herring which we would eat in comfort on our beds. So far we had not been caught. One Saturday, however, was our Atonement Day. Outside, a blizzard was howling. We were sitting on our beds, waiting for supper, when suddenly a high-pitched, nervous roar sounded from the door: "Ninth Battery, alarm!"

We recognized Kalugin's voice and jumped to our feet. He was standing in the doorway, his nervous, dry face contorted with concealed excitement. This did not suggest anything good. Fixing our belts, we were about to put on our coats when we heard him laugh maliciously.

"Get out onto the yard without coats!" he screamed. "Fast!"

"Well, well," moaned Lolek, running by my side, "this is the last straw!"

Standing motionless in the yard, shaking with cold and fear, we watched our boys returning from the mess hall with the tins filled with our bread, herring, and tea. Kalugin gave them a dirty look.

On the corner there was a sewer. The group were about to pass it by, but seeing Kalugin and the battery, stopped in silence. Kalugin ordered the first one to lift the sewer bars, throw the contents of all the mess tins down the sewer, then join the ranks. We stared at our bread and herring as they went tumbling down the black hole. Suddenly, out of the blue, Shaposhkin appeared. On the *kombatr's* orders he took down the names of all those present to make sure no one was going to miss the fun. Then Kalugin, his face still contorted, began cursing us in a broken voice. He had thought, he stuttered, that we were well-disciplined and good soldiers, while in fact we were a band of rascals, disregarding orders, a bunch of women afraid of cold weather. The "rottenness of the West" was coming out in us, but he, Kalugin, would let it go no further, and so on. It was terribly cold. The lieutenant let his furious eyes rove over us all. He was undoubtedly a little drunk. He noticed that Jan, standing opposite him, was rubbing his frozen ear.

"So you are freezing, you sons of bitches?" he howled, his voice breaking. "Right turn! Double march! Run!"

We moved off, not knowing what to expect. Kalugin ran behind the last row of four, scaring us with that sharp voice that did not admit opposition. We ran from the yard into the drill fields, keeping close to the wall. Having made a round of the field once—a good kilometer—we began gasping for breath. And all the time Kalugin kept shouting: "Faster!" Along the wall the snow was deep, making running difficult. In spite of the cold, drops of sweat fell from our foreheads. We sensed that this game was far from its end. In the middle of the circle we tried to slow down. But Kalugin roared: "Faster, sons of bitches!"

He himself was running easily. The anger and the vodka had not yet evaporated from him. Somebody in the rear ranks began to slow down. Then we heard the first pistol shot. I turned around in terror. Kalugin was running ten meters behind the last row, firing two *nagans* under the feet of those in front of him. Neighing with mad joy, he was loading his guns on the run and firing them off like a man possessed by the devil. We had made the round more than three times when Joziu a frail, small youth, unable to stand the speed, tripped and fell. Kalugin instantly recovered his sanity. Ordering the battery to halt, he ran to the boy and with great concern helped him up from the snow. Softly, almost timidly, without looking anyone in the face, he ordered us back to the barracks. We moved off in silence, panting from the effort and excitement. In front of the barracks we found a crowd of soldiers, alarmed by the gunshots. An officer on duty came running toward us, wanting to know what had happened. Nothing had happened, Ludwik told him. Our *kombatr* had simply instructed us to practice firing *nagans* at night. With a surprised face, the officer turned and went away.

"Boys," said Ludwik, "I hope this will be kept to ourselves. If the major gets wind of it, our Kalugin will get kicked out. We have no right to complain about him. He's better than all the other officers."

Everybody agreed that Kalugin had to be saved from the consequences of his drunken fit. A man with no family, on bad terms with his fellow officers, he spent most of his free time lying on one of our beds, listening eagerly to stories of life abroad. So when we saw him walking slowly over the field, all alone, with his head down, we felt we had to tell him to forget the whole affair. Ludwik, sensing that we were about to ask him to do the talking, slowly went up to Kalugin. They both stopped. We did not hear the conversation, but we saw how the *kombatr* slowly straightened his posture, how his expression changed, how he finally smiled and solemnly hugged Ludwik and kissed him on both cheeks. Then he straightened his cap and walked quickly toward the barracks. As he passed us we saluted. Returning the salute, he walked on a few meters, then stopped, turned around, saluted again, and said simply:

"Thanks, boys."

Next day, Kalugin, instead of drill, treated us to an "educating" discourse on the destructive results of alcoholism. We listened attentively, without revealing that we understood his intention of justifying himself in our eyes and thus punishing himself.

But as a result of that last run in the snow without a coat, little Józiu fell ill. When we took him to the sick ward he had very high fever. Next day, he was seriously ill with pneumonia. Kalugin sat by his bed four nights in a row; he neither slept nor ate, but kept himself going with cigarettes. There was great fear that Józiu would die, but, thanks largely to Kalugin's devotion, the boy began to regain health and eventually recovered.

Though I, thank heaven, managed to escape taking part in any of the famous twenty-four-kilometer runs, I nevertheless happened to get involved in a result of one of them. This run had been particularly unpleasant because it took place in a blizzard, in bitter cold and darkness, and the runners had to practice sharpshooting at targets while they stumbled through the snow. During this ordeal Julek lost one shell, and Stefan two. It is surprising that under these conditions only three were lost.

But Kalugin and the other officers had a different view of the matter. At dawn the next day they chased us out onto the route of the run, simply to make us search for those godforsaken shells. The task was absurd. The snow came up to our knees. The route had been tramped on by thousands of feet. We had to get down on all fours in the snow, where we moved about like animals, cursing our luck, the army, and above all the shells. Long after sundown, exhausted and frozen, we returned to the barracks for supper without the shells. The second and third days the story was repeated. With shovels we dug a whole square kilometer of snow without any result. Lolek morbidly predicted that this would be our occupation until the snow melted in the spring. On the third day Turadov, a Turkestanian driver and friend of Aram's brought us dinner. After looking carefully around, he handed me a little package. In it were three shells.

"From Aram," he whispered. "Continue your search until the evening so that no one will suspect that I brought you these."

I nodded, put the shells under my cap, and ate my borscht in peace. We were saved. I distributed the shells, one to Julek and two to Stefan. In the evening they reported to the *kombatr* with the "found" shells. Kalugin was clearly puzzled. Though he did not say anything, his dry face revealed relief. In the barracks, when I tried to thank Aram in the name of the whole battery, he interrupted me immediately.

"One does things for one's friend," said he, "but one does not expect thanks."

★ ★ ★

ON Christmas day we naturally had our normal occupations; but in the evening we were allowed time to celebrate the season of the year. Almost everyone had a package from home with the traditional Christmas wafer and home specialties. We—Poles, Ukrainians, Jews—all shared these wafers, which reminded us, not of the religious significance of Christmas, but rather of home and Poland. Kalugin, who knew something about the holidays and understood their significance, but looked at everything with an unfavorable eye because of his decidedly "programmed" atheism, shared a wafer with everyone and listened to the carols and stories about Christmas traditions.

On New Year's Eve a great surprise awaited us. Around 10:30 a few wooden boxes were carried into our dormitory. The soldiers who brought them informed us that they had been sent by Captain Kovalski. We were amazed. We opened the boxes. On top of each lay a card with an inscription in Polish: "Happy New Year from a countryman." The boxes contained bottles of red wine, gingerbread, a few rings of sausage. How Kovalski obtained these treasures we never discovered. He turned up in the dormitory just before midnight. When we started to thank him, he was almost indignant. He sat on a bed in the middle of the room, and we drank to a prosperous New Year. Then he began to talk about himself. For the first time he declared his "Polishness." He was an engineer from Warsaw. When he was twenty-five-years old and a burning "Red," he had taken part in illegal communist agitations in the factory where he worked. When the police discovered his activities he fled to Soviet Russia with his wife, a girl of twenty. Because of insufficient references from the Communist party, both he and his wife had spent the first year—1935—in prison. In 1936, on the intervention of some party big shot, they were freed. The wife, unfortunately, had contracted tuberculosis in prison and died four months after her release. When speaking of her, Kovalski's eyes hardened. Because of his qualifications, he had then joined the army as an engineer. By 1938 he had reached the rank of captain. In September 1939, just before the campaign against Poland, he was arrested a second time. It was feared, as he said, that his patriotic feelings for his former fatherland would get the better of him and lead him into conflicts with his duties. At the beginning of November, before the Finnish campaign, they liberated him, apologizing solemnly and explaining that there had been a mistake. He was decorated twice on the Finnish front. He told us all this with an indifferent, expressionless face—the result, we felt, of routine and a few years in Russia. Then he changed his topic and began to inquire about Warsaw under German occupation. In answer, somebody very thoughtlessly said:

"Oh, there it's even worse than in Lwów."

A hardly noticeable smile crossed Kovalski's face. He pretended he had not heard. Dropping that subject, he began telling us about life in Russia. He spoke coldly and reasonably, without sentimentality. He enumerated the difficulties that still confronted the government in their attempt to achieve prosperity and happiness for all citizens. He talked about the five-year plans, their successes and failures. There was a great deal of truth in his explanations, but he carefully omitted such topics as that of democratic freedom in this "fatherland of the proletariat," where such freedoms, supposedly developed to the highest levels, did not even exist. He failed to mention that in 1939 more than twenty million people were in prisons, concentration camps, and forced labor camps; that when men wanted to discuss the regime, they lowered their voices and looked around with fear; or that even the smallest and often absolutely unjustified denunciation caused people to go to prison for many years.

On these topics Kovalski was silent. As if he had read our minds, he began to explain that if the work of introducing the idea of communism had started in Europe—in England or in Czechoslovakia, for instance—it could have achieved splendid results in a far shorter time and without the difficulties and sacrifices that the USSR had to suffer. He reminded us of the condition Russia was in after the first World War, of the intellectual level of society, the ignorance and illiteracy throughout the Russian countryside.

In comparison, he emphasized the tremendous progress that had been made in every respect during the last twenty years.

We were very polite. Nobody asked any questions. He then passed on to the subject of distribution of power among the nations of Europe. We listened eagerly to what he would say about Germany. And then for the first time in Russia we heard somebody admit that the alliance with Germany could not last long, that soon it would have to come to a conflict. National Socialist Germany was now in a position to conquer all of Europe. If she achieved that, then she would be bound to turn against Soviet Russia—a country Germany hated, if only because of ideological differences.

"Well, what's Russia waiting for?" someone asked. "For the day when Germany, possessing all Europe and consequently all . its industry, turns against Russia and defeats it with one blow?"

Kovalski raised his head. No, it was not that. Russia's present leaders look into the future; they are well aware of the inevitability of the conflict. But they also know that today Russia is still far from ready to fight, and if Russia is sending wheat, gasoline, and food to Germany, it's not from any love of that country, but to mislead its vigilance and to gain time. Before long, the conquered and persecuted peoples of Europe would be full of hatred toward the Germans. When war began between the two colossi, these people would be organized and offer Russia their help. Even Poland, which had no sympathy with Russia and its regime, would join Russia against the most abhorred enemy of all, Germany.

Again there was a great deal of truth in what Kovalski said—mostly in what he said about Russia's lack of preparation for war.

No one more than ourselves in Europe at that time had had a better opportunity to appreciate that lack. We remembered the endless squadrons of airplanes, the powerful mechanized columns of the Germans that had swept through Poland in September as if they were just holding drills. We remembered their splendid armored automobiles, motorcycles and sidecars, equipped with machine guns—everything made of materials of the finest quality, carefully worked out for the comfort of the soldier.

Against all this we tried to oppose what we had seen of the military force of Russia. At first, it's true, Russian tanks did inspire confidence by their appearance and great quantity, but in no time we'd seen them abandoned in great numbers on the roads because of their defects. Then there were the airplanes, powerful four-motor bombers which the Russians themselves admitted were no good as fighting planes; the *istrebityele* or "destroyers"—swift and handy, but without armor and with very small firing power. Then the infantry, a mass of badly uniformed, badly booted men, whose guns hung from their shoulders by pieces of string! Even the bayonets, while on the march, were tied by rope to the rifles. Supplies, in great part, came in horse-drawn, flat-bottomed wagons, the harness made of straps of coarse cloth instead of leather. As for the faces of the masses, particularly those of the infantry, of whom a large percentage came from the East, they looked wild, stupid, without a shadow of intelligence. Nor did the officers reveal, as a rule, any great spiritual values.

Now, being ourselves in that army, we were naturally able to observe much more than before. It was certain that the morale was very unsatisfactory. I did not meet anyone, except perhaps the *politruk* or his aide, *pompolit*, who did not talk against the regime. This often made us wonder how those with whom we talked would behave in the event of war with Germany. Would they fight, concealing their indifference or hatred for the regime, or would their patriotic sentiment, that love for "Mother Russia," get the better of them?

Such were the questions we asked ourselves as we sat with Kovalski during the last hours of 1940 and the first of 1941. Around three o'clock the captain once more raised a toast for a Happy New Year, said good night very affectionately, and left.

★ ★ ★

DURING the first days of January came the most unexpected news that we were to leave Odessa for a "polygon" [artillery range] beyond Dniepropietrovsk, somewhere near Novomoskovsk. Preparations were surprisingly short. In two days we were already loaded on a freight train bound for Dniepropietrovsk.

The farther we traveled from Odessa and the Black Sea coast, the deeper grew the snow. Winter was at its height, the thermometer dropping as low as twenty-five degrees below zero. In the center of each car on the train stood a little iron stove, its pipe let out at the ceiling. We rode forty men to a car, just as we had three and a half months ago, and how everything had changed since then! Now we were old soldiers; all our features had hardened and grown coarse. Our habits had become more vulgar, and so had our language, into which we often interpolated Russian words. We no longer covered the table with newspapers when eating; the little courtesies we had shown one another had also disappeared. This was not due to selfishness or lack of cooperation, but was a result of the habit of having to do everything for oneself and of expecting the same from others. The one who suffered most from this was Jan, who still appeared with a lighter when anyone rolled a cigarette. He was constantly repeating:

"Gentlemen, don't let's forget we are Europeans."

We agreed with him, but nobody felt like imitating his efforts to conserve good manners and habits.

After three days we reached Novomoskovsk. Because of sharp frost and heavy snow we were not to build little huts on the ground, as had been announced, but to find billets for ourselves in the village of Orlovaya, not far from one side of the polygon. Jan, Julek, Ludwik, the junior *komandir*, Vania, and I were quartered together. Our hut looked decent, and its proprietors quite kind. We were given a room next to the kitchen, and therefore well heated. In the room there was just one object—a wooden bed. After very careful observation, we generously gave the bed to Vania. It was so full of bedbugs that it seemed to be moving by itself. The rest of us were given two straw mattresses which we placed on the ground (the hut had no floor) as near as possible to the kitchen stove. Covering these with our blankets, we felt very satisfied with our new quarters.

On the very first night I had a peculiar adventure. The sentry guarding our guns, which were parked outside the village, was from our battery. I, who in the garrison always managed to avoid guard service, usually by explaining that I had regular work at the parking lot, was now going to have to do it, without any doubt. The guardhouse was some eight hundred meters from the guns. The chief of the guard, a junior staff lieutenant too big for his boots, gave us a long speech about how we'd have to be constantly on our toes, as the region was very dangerous and attacks on sentries quite common. On the slightest suspicion—shoot! Well, for my first effort at guard duty, I thought, the prospects are rather terrifying! My service was scheduled for eleven o'clock at night. Just before that hour, the officer in charge led me through deep snow to my post. There I relieved Lolek, who returned with the officer. I was left alone in the darkness. All I could see, silhouetted against the snow, were the contours of ten huge guns and tractors. To save their tires, the guns had been raised and placed on wooden blocks. Slowly my eyes became accustomed to the darkness, and I began to distinguish trees and an enclosure on my left, opposite a group of shrubs. As I wandered past the guns, holding my rifle at the ready, I felt very uncomfortable. At one point, while my back was turned to the guns, I suddenly heard a loud report, like the crack of a breaking branch. I turned and, according to regulations, fell flat on the snow, where I assumed the position required for shooting. My heart was beating like a hammer. I gazed straight ahead. Suddenly, some 150 meters away, I saw very distinctly a human head, motionless in the snow. Its hair was dark, covered by a cap of a lighter color, probably a fur cap. I swallowed once in great fear, and then, taking careful aim at the head, I screamed: "Halt! Who goes there?"

Not a sound. The head didn't move. I called once more, and then a third time, warning the head's owner that I would use my weapon. No answer. No doubt he is counting on the fact, I said to myself, that I cannot see him properly. Then, with great precision, I aimed—and pulled the trigger. The noise of the shot came back to me in an echo from the wood. The head still hadn't moved. Then I heard voices coming from the guardhouse, and the tramping of feet. As they approached, without taking my eyes from my sights (all this according to regulations!), I raised my left hand. I heard the words of the lieutenant's command. Then the entire guard made a dash towards me, the lieutenant collapsing on the snow beside me. Without speaking, I indicated the head in front of me. The lieutenant looked and cursed:

"Son of a bitch! Did you kill him?"

"I don't know," said I.

After a while the guard, divided in two groups, encircled the "son of a bitch" from two sides in two broad arcs. The lieutenant and I remained on the snow—I still aiming my rifle, he squeezing the *nagan* in his hand. In front of us, from both sides, the soldiers threw themselves onto the head in the snow and—a great howl of laughter resounded. They went on roaring as if somebody were tickling their feet. The lieutenant and I looked at each other. I remained where I was but stood up, while he ran forward to the men. The human head turned out to be a tree trunk covered with a thick cap of snow. They laughed, while I cursed. The lieutenant was silent for a while. Finally he turned to me:

"Return the shell!"

I was numb with fear. The shell! Of course, I hadn't thought of that. I began feverishly looking around in the snow. No use. Naturally, there was no trace of it. The deep snow, tramped by many feet, had devoured the shell the way it had devoured those famous three shells after the twenty-four-kilometer run. The young lieutenant (he had just arrived from an officers' school) could now be important again.

"You will be held responsible twice," said he. "For unnecessary firing and for losing the shell."

We heard a scream from the direction of the tree trunk. The men were still there, still laughing wholeheartedly. I saw Lolek shape his hands in a trumpet as he yelled:

"The bullet is bang in the middle of the trunk! Perfect aim!" We approached the tree. In the center of the trunk the bark had been torn off, revealing a hole in which you could feel with a pencil the steel of the bullet. We gave up searching for the shell. I didn't care. I'd have to go to prison anyway, for "unnecessary firing." They'd probably add an extra three days for the shell.

The following day we came off guard at three in the afternoon. I went back to the battery district. The story of my shot was already well known throughout the regiment. At five o'clock Kalugin told me to report to Captain Lermontov, who was replacing the regimental commander at the polygon.

"You shoot very well," smiled Kalugin, "but now you're for the guardhouse."

Lermontov listened to my report without lifting his head. Finally he stopped writing and looked at me.

"For using a weapon without real need, I give you six days of strict arrest," he said.

I mumbled, "Yes," saluted, and waited to be dismissed. But Lermontov smiled lightly and continued:

"But for your excellent aim and quick decision I cancel the punishment. Besides, tomorrow you've got to go to Odessa. They need drivers there for some long trips. Before you leave, come and see me and I'll give you a letter for Irina."

Well, I was lucky that time. I left the room, surprised and startled. I went back to Kalugin and reported everything to him. He wagged his head and said:

"Well, with so much pull you must be very good to Lermontov's sister!"

I returned to my quarters. They boys were delighted with Lermontov.

In bed that night I began wondering about the destination of my "long trip." Maybe to Poland? Then I fell asleep.

★ ★ ★

NEXT morning I was overjoyed to hear that Aram was to accompany me to Odessa. The journey took two days, during which he and I and two others shared a loaf of bread, a piece of bacon, and a can of kasha which we couldn't cook as we had nothing to cook on. By the time we reached our barracks we were all half-starved. After a proper meal, we retired at once to our own beds, too exhausted to look for our comrades.

The following day we discovered where the "long trip" was to take us: the regiment was soon to be transferred to a garrison in Bessarabia, to territory annexed by Russia from Rumania in the summer of 1940. Our detachment was to be sent ahead to prepare the quarters of the evacuating regiment for our regiment's arrival. As usual, our exact destination was a mystery. The commander of our billeting division was a certain Lieutenant Yarmultovski, an exceptionally handsome boy just graduated from an officers' training school. The second-in-command was Budanov, a short man who looked like a monkey and was supposed to know something about automobiles. The noncoms consisted of an imbecile of a sergeant-major called Riklin; Stiopka, a pleasant boy; and the guitarist, Andriusha. We were to be given seven ton-and-a-half trucks, whose drivers, apart from Aram and myself, included the huge-headed Dmitrov, a perfect colleague and companion, and Shvetzov, a disagreeable peasant with a dark complexion and a similar character. At the last moment, and to our delight, we learned that Misku, the tile-maker whom I'd known in the H.Q. Battery, was also to join us.

While waiting for the whole detachment to get organized, I officially and with the full knowledge of Yarmultovski, moved to Irina's house. I was very comfortable there. Most of my time I spent thinning out the Lermontov's well-stocked larder. I slept in the captain's silk pajamas (from Finland, of course) and smoked his cigarettes, of which he had large supplies. All in all, I spent those few days as if I were taking a fattening cure. Irina was actually quite nice; she did not make her usual scenes, nor was she particularly demanding.

At last they called me. Our trucks were being loaded, mine with a supply of gasoline, oil, and grease; another with an unfortunately small store of food, all "dry rations"; and a third with a load of uniforms, including a change of boots, underwear, straw mattresses, sheets, etc. The remaining trucks contained tools. We moved out through the gate. After a mere half kilometer the first truck broke down from engine trouble. It was a little thing, but to repair it took a whole hour. One more kilometer and Aram's truck went on strike; then my carburetor blew up. On each occasion Budanov, the technician, did his best to help, but alas, the poor man didn't know much about his job. We didn't blame him, but kept wondering what would happen if all the trucks should develop trouble. We had started off at eight o'clock and by noon had covered two whole kilometers. From then on, however, we rolled along pretty well.

Unfortunately I'm rather hazy about the route we took. I know that in the early evening, not far from Odessa, we crossed a long bridge over a part of the Dniester Bay. Then we came out onto the recently annexed territory of Bessarabia. Snow had been falling heavily since the early morning. None of us, of course, had any chains for the tires. Soon after crossing the Dniester bridge we came to a steepish hill where Dmitrov's truck, which was heading our procession, could go no further. Passing the three trucks ahead of me, I began with great difficulty to climb the steep incline, the rest of the trucks following on my tracks. In this way we managed to mount the dangerous hill, and moved on slowly, covering an other ten kilometers. The snow kept falling, while a high wind blowing across the road piled avalanches of snow on its sides. The men on top of the trucks froze terribly. Presently the road came to a sudden end. A great mountain of snow reared itself up in front of us. It was impossible to pass and impossible, in spite of our efforts, to shovel a passage through. There was nothing to be done but seek some human habitation where we could spend the night and get help to clear the road when the storm was over. We sent out patrols to look for a community. For the next two hours we just stood there freezing, watching the trucks slowly but surely disappearing under the snow. Then, out of the darkness, Aram appeared. (Who but Aram could have found human habitation in that hell?) About one kilometer from here, hidden in a valley, he said, lay a Rumanian village. We moved on in single file, holding onto one another so as not to get lost. The trucks remained on the road, sacrificed to fortune. We just let out the water so that they would not freeze up.

After stumbling through deep snow for some time we noticed, down in the valley, a few weak lights flickering from tiny windows. Yarmultovski, who was leading, stopped and waited for the rest of us to join him.

"I'm going to the first hut," he shouted into the wind. "You billet yourselves wherever you can. Tomorrow we'll have an assembly in my hut. Maybe we can leave in the evening."

"Evening! Evening!" muttered Aram. "We won't be able to dig the trucks out till spring comes!"

Aram, Misku, and I set off into the village. Aram didn't make for the first hut. He was choosy. Picking out a stone-built, solid cottage, he knocked at the door. It was opened by an old man with a gray beard. He was dressed in trousers and shirt of homemade cloth, both white and clean, and he was barefoot. At sight of our uniforms he looked surprised and afraid. Nevertheless he opened the door wide, and with a gesture invited us in. Beyond the threshold a group of children stood staring at us. A small boy brought us straw slippers and suggested we take off our boots. Then we told the old man what had brought us here, that Misku and I were Poles and Aram an Armenian. The old man nodded and asked us to come in. His Russian was not much better than ours. We found ourselves sitting down to table with the family. This family numbered at least fifteen, ranging in age from three to a hundred years, and included two young married couples.

The supper was strange. The old man made a sign of the cross over a loaf of fresh black bread from which he proceeded to cut tremendous slices, first for ourselves and then for the family. In the center of the table stood an immense bowl in which lumps of lamb, potatoes, pickles, and tomatoes floated in boiling oil. While we dipped into this bowl with a wooden spoon, a boy with a large earthenware pitcher kept pouring sweet red wine into our mugs. The women, I noticed, did not drink wine. During this meal, which lasted a good half hour, no one spoke.

    The room we ate in was the center of the house. Against its longest wall stood a typical Russian stove, the so-called *pietchka*. This stove had a number of different levels, niches, and terraces, all the way up to the room's high ceiling. The whole family slept on these terraces and niches. You might say that they lived on this stove. Here they were conceived and born, and here they died.

    After supper we chatted with the old man for a little while longer; then he made the sign of the cross over the table and we got up. Our beds were on the stove's most honored recesses, not too high, not too low. The bed "linen" consisted of coarse Rumanian rugs and smelly sheepskins. The big oil lamp which hung over the table was put out. But a small red flame, in a corner among icons and sacred pictures, remained burning. Without any embarrassment the whole family began to undress, then they climbed on the stove to occupy their usual places. Despite my fatigue I could not fall asleep. The young, powerful wine kept coursing through my head; the room was very warm and the odor none too delicate. Above me, one of the young couples kept giggling and making unseemly noises. Aram, lying beside me, spat over my head in a big arc.

    "And they call us barbarous people of the East!" he muttered. "In my country such customs, such behavior, would be impossible to imagine!"

    When we woke in the morning there was already a great commotion in the hut. Only the old head of the family was still lying on the stove. We again received a strange meal—boiled paste in boiling milk. The snow continued falling heavily. We had not the slightest intention of looking for the lieutenant; an "assembly" would have been nonsense. Next to our cottage we discovered a small rural store—pretty well cleaned out, of course, since this summer of "liberation," but you could still get cigarettes, tobacco, matches, sugar, and what was even stranger, chocolate—old chocolate, which had probably been doing penance here for many years. Aram and I found everything indecently cheap.

    The peasant's attitude toward us was friendly but cautious. To this distant rural community the Russians had not yet penetrated, so that the native didn't know what they were like nor how he should behave toward them. Aram, biting on a piece of moldy chocolate, was smoking the quite good Rumanian cigarettes and looking forward to the heavenly delights of Rumania.

"If a village like this has so much of everything...," he reasoned.

Here he was mistaken. In bigger towns there were a lot of "capitalistic riches," but the Russians had bought almost everything, and what they hadn't bought was so expensive that there remained very little we could afford.

Just before dusk, as the storm began to weaken, we found the house where Yarmultovski and Budanov had stationed themselves. They had been given a large room, covered with rugs, into which the whole family kept running to serve the "powerful chiefs" of the Russian Army. The lieutenant, resting on a heap of skins, was smoking a cigarette with an air of great satisfaction.

"When we'll start?" he repeated with surprise, in answer to my question. "Not till it stops snowing. And even then we'll have to wait till the roads are clearer. Why?" he asked with a smile. "Aren't you happy here?"

In the doorway we ran into the imbecile, Riklin. He was laughing broadly, showing his black teeth. He explained to us that the lieutenant and Budanov were being received like princes, that they had even taken advantage of two girls of the house.

"I also slept with one," he boasted.

We went out into the street.

"If this continues," observed Misku, "we really will, as Aram said, be spending the winter here, but for different reasons."

We were not particularly worried about that. We were warm, we ate a lot, and we had nothing to do. We were in no hurry to reach our destination. Our whole division, quartered in the huts of the village, was very comfortable; the majority even enjoyed the same luxuries as the two officers and Riklin. The Moldavian village's ideas of hospitality were very broad. Misku and I, however, bearing in mind the reputation of Rumanian rural communities for venereal disease, lived virtuous lives.

Despite the fact that it was so close to the main highway, we were stuck in this village, cut off from the world, for two weeks. Even when it stopped snowing the road was still untouched. One day we all set off in the direction of our abandoned trucks. We found the highway by means of its telegraph poles, but the trucks were invisible. We searched for half an hour before

Shvetzov hit something hard with a pick; it turned out to be the fender of his own truck. Even with the help of the peasants it took us a whole day to dig out all the trucks. There was no question of moving them. After a conference, Budanov and one of the noncoms borrowed a sled from the peasants and set off to a neighboring village to telegraph the military authorities in Benderi. Next day they returned with the news that we were to wait until the army and civilian population had cleared the road. So we stayed on in the hospitable village, our one pointless duty being to stand guard over the trucks!

At last a motor plow and swarms of peasants with shovels arrived from Benderi. Within a few hours the road was cleared; the motors were put in running order and the whole village turned out to witness our departure. While the girls, concubines of many soldiers, sobbed, our old host, the village mayor, made a long speech in broken Russian to the lieutenant. Yarmultovski offered his thanks in our name, shook hands, and off we went.

After Benderi, we passed through Akkerman, and eight kilometers further reached what turned out to be our destination. It was a small settlement called Mansburg, which had been colonized by Germans and only recently abandoned by them on the basis of exchange of population with Germany. The village they left behind contrasted strangely with the miserable Bessarabian villages. Mansburg was built after the German fashion. The cottages were decent, spacious, light, and airy, each surrounded by a small garden. The little church was built entirely of stone. The stables and outhouses were large, with brick floors. Every building was roofed with tin or tile. On the outskirts of this exemplary village you could see the natives' low, wooden, dirty hovels, usually poorly thatched, and behind them filthy back yards.

Until recently another artillery regiment had been stationed in this settlement. Groups of men were still liquidating their quarters. We occupied four cottages on the edge of the village, opposite the Protestant church, where the previous regiment had stored spare parts for their trucks. The following day, with a fury typical of Russians when complying with duties, and a sudden ardor due to four weeks of unnecessary delay, the "commission" proceeded to inspect the regiment's billets. The cottages, after six months of our predecessors, presented a desperate and desolate sight. The stoves, Misku's kingdom, had fallen apart; the windows were either broken or had been simply taken out of their frames, the doors torn from their hinges. In what had been the officers' dining room, the mirrors were smashed, candlesticks destroyed, and the walls pocked with bullet holes.

We began feverishly putting these billets into some kind of order. A few days later, however, Yarmultovski received a telegram from our major informing him that, instead of moving to Mansburg, the regiment would most probably stay in Odessa, and that until a decision had been reached we should remain where we were. Yarmultovski and Budanov then quietly left for Kishiniev in order, as they put it, "to inform ourselves about our future movements." The remnants of the previous regiment also cleared out. Up to this time we had been eating in their mess, but now that they were gone we were supposed to get our food from the military stores in Akkerman and cook it ourselves.

Now began the strangest story of my military career. Since the officers had left, Riklin and I went off to Akkerman to fetch the food. The head of the storehouse, a noncom, listened to Riklin's report, looked at the slip of paper signed by Yarmultovski giving the number of men in our group, wagged his head, and said he could not hand out the supplies because he hadn't received the order. All explanations failed. When Riklin reported our situation to the C.O. at Akkerman, he shrugged his shoulders. He had received no order about fifty men of any Odessan regiment. It was our officers' job, he said, to come to him personally. We returned empty-handed.

On the third day Riklin and Shvetzov left for Kishinev to look for our officers, taking along some of our extra uniforms for sale. They had to live on something. Meanwhile, we were hungry. In fact, we were pretty near starvation. Everyone fed himself on his own. Aram, Misku, and I happened to find half a bag of potatoes in an abandoned cellar. In Akkerman Aram sold his spare tire for a few kilograms of sausage and bacon. The other managed in a similar way, selling uniforms and other government goods.

But twelve days passed before Riklin and Shvetzov returned—without food, without the uniform supply, and without a cent; but so full of vodka that you could smell them a mile off. They hadn't seen a sign of either Yarmultovski or Budanov. It was quite clear that in order to exist we'd have to rely on our own wits. Misku was furious.

"Now's the time to desert," he growled. "We have every right to do so."

During the next three days we sold every scrap of our supplies, dividing the money equally between us. There wasn't much. Aram then found a buyer for a barrel of gasoline which happened to have been loaded on my truck. The situation was very sad.

It had grown no less sad when Andriusha rushed in one evening in great excitement. For some time now he had been wandering off to a recently established cooperative farm (kolkhoz), where families imported from Russia had begun to cultivate the land abandoned by the German colonizers. In addition to the kolkhoz, there was an M.T.S. (Motor-Tractor Station) which rented to the kolkhoz the mechanized equipment for the soil's cultivation. Strangely enough, the M.T.S. service employed only two mechanics, and the kolkhoz itself had only eight men and some forty women! And among them not one tractor driver! They were in dire need of about fifty people to begin the job of plowing and sowing. Andriusha had had a talk with the chief of the kolkhoz, to whom he had offered our help in return for food and a little pay. The chief was delighted.

Next day we marched the three kilometers to the kolkhoz and set to work. I drove a tractor, pulling plows, harrows, and other farming machines mysterious to me. A large, tasty breakfast was served us in the dining room of the kolkhoz. Dinner was brought out to the fields, and for supper we returned to the kolkhoz. We were all very pleased with ourselves, and beginning to like our new life and occupation. Meanwhile, a thaw had melted the snow and started floods, so that even if we'd received an order to return to Odessa the roads were in too bad a state to have carried it out. Some of the soldiers established themselves for good in the building of the kolkhoz. Free love flourished. The women, short of men, were delighted with this sudden help. (The men supposed to have come to this kolkhoz from the Ukraine had never turned up.) To the admiration of the whole community, Misku constructed a big tile stove in the dining room. This stove earned him great admiration and respect, not to mention the eternal love of the cook, who, though certainly in her forties, had the vitality of a teen-age girl.

We worked merrily, but hard, from morning to night. Because of the spring weather nobody wore a uniform or even a cap. In fact, we showed no signs of being members of an army.

Suddenly, at noon on April 20, after three weeks' absence, Yarmultovski and Budanov returned. In our billets, of course, they didn't find a soul. In the evening I accompanied the few men who still slept in the huts. We wandered into the village dressed, as usual, in our shirt sleeves. There on a bench in front of one of the huts, sat the lieutenant and Budanov. We stood still. The lieutenant, in the glory of his uniform, properly buttoned and belted, rose with a severe and ceremonious face.

"What is this?" he roared. "Are you deserters or Red Army men?"

We stared at him, speechless, and entered the hut. He followed us and, trembling with indignation, howled again:

"What? Mutiny?"

Dmitrov, a quiet man, calmed him.

"Not yet," he muttered.

The lieutenant relaxed.

"Where are the men?" he asked.

"Well," said Dmitrov peacefully, "they are working at the kolkhoz for their daily bread. It's either that or starvation."

Yarmultovski stared at him in amazement.

"What's that you're saying?" he croaked.

"Exactly what you hear," Dmitrov assured him.

Then we began to explain what had happened since they left. Tearing his hair, the lieutenant raced back and forth like a lion in its cage, swearing he had no idea anything like this could have happened. He imagined we'd all be so pleased with our freedom. Budanov, however, was silent. He just sat, his eyes blank, paralyzed with fear. Then, with a face void of expression, he spoke.

"We shall be court-martialed for this!"

Yarmultovski let out a howl, threw himself on a bed, and began to sob. We were embarrassed. After all, these two fools had not meant to cause all this trouble.

Dmitrov went off to the kolkhoz to summon the rest of our gang.

Yarmultovski grew a little calmer. He kept repeating some remarks we had made about the sale of the uniforms, gasoline, etc.

"Maybe we can fix it somehow," he said, more to himself than to us.

Budanov, however, was silent. He just sat, staring stupidly at nothing, smoking one cigarette after another.

At last the rest of the gang arrived. The negotiations, at first, were stormy, reminding us of a scene on a ship whose sailors have mutinied against their fool of a captain. But very soon our "captain" changed his tactics. He was quiet, full of humility, submitting to all the reproaches leveled against him. Then he had another idea. Yes, he and Budanov would be court-martialed, but so would we, for selling state property, for appearing in public without uniforms, and for working at the kolkhoz. This might have been necessary to avoid starvation, but who would take that into consideration? They'd simply tell us we should have gone to Akkerman and given ourselves over to the town major, who'd have fed us until the matter had been cleared up. At this, Aram, Shvetzov, even Riklin, were indignant.

"The son of a bitch," growled Shvetzov, "now he's trying to threaten us!"

The majority, however, agreed with the lieutenant. The affair might end as badly for us as for the officers. Dmitrov was the mediator.

"What is your suggestion, *tovarisch* lieutenant?" he finally asked.

Yarmultovski lifted his head.

"Mutually to forget each other's sins," said he seriously. "You ours, and we yours. Don't you mention a thing that occurred here, and I will fix the problem of the sale of the supplies. I will also forget," he added, "what you all look like at this moment!"

After a stormy discussion we decided to accept the conditions. The lieutenant sighed deeply. We also felt better. Budanov, however, was silent. He just sat, his eyes blank, apparently overwhelmed by the proceedings which in some mysterious way were to prevent him from being court-martialed.

★ ★ ★

NEXT day Dmitrov and the lieutenant went to Akkerman to inquire about road conditions all the way to Odessa. We had decided we couldn't return to Akkerman and once more ask for food, since this would create suspicions as to how we'd been feeding ourselves in the meantime. So there was nothing for it but to work in the kolkhoz to the last day.

The lieutenant returned with the news that the road beyond Akkerman was absolutely hopeless. We'd have had to wait until the floods had gone, and that might take a month. We could have stayed peacefully in the village, but Yarmultovski was impatient. His one desire was to put an end to this strange escapade and return to the regiment as soon as possible. So he began trying to get permission to load us all on a train. This wasn't so easy; it was five days before he managed to obtain the required number of freight cars.

We said good-by to the kolkhoz, to the once German colony, and we went away. In Akkerman, the trucks were loaded onto seven open trolleys, and ourselves into a freight car. The latter, however, was so crowded that I preferred to remain in my truck on the trolley, where Misku joined me. During the night we established ourselves on top of the truck and went to sleep under the calm April sky.

★ ★ ★

ON rejoining the regiment, the first news we heard was that Kalugin had been transferred from our battery to regimental H.Q. In his place, Lieutenant Karpyenko became our kombatr. Karpyenko was undoubtedly a talented officer; he held in high esteem the combat efficiency of our battery, whose marksmanship in the winter school area had impressed the entire Odessan garrison. Unlike Kalugin, however, he stuck very close to army regulations, permitted neither himself nor others any deviation from the rules, was extremely demanding, and had absolutely no sense of humor.

On the heels of this news came more, good and bad. First, on May Day, in appreciation of the results of our training, we privates were to receive the titular commission of "junior sergeants"; and second, that the attitude toward us *zapadniki* was to be made more rigorous. According to this order—which was unofficial—conversation in Polish during working hours and in the presence of superiors was to be prohibited. This prohibition was contrary to all the provisions of the Constitution, which, among other privileges, claims freedom of language. An immediate consequence of this order was a deterioration in the relationship between the men and their superiors. The former—including even Ukrainians, who had always spoken to one another in their own language—now began to converse pedantically in Polish.

The joy, therefore, of finding myself back among my old friends was considerably dampened by the new situation, which everyone looked upon as only the beginning of further chicanery.

I reported to Karpyenko, who received me in a friendly way. After inquiring about my impressions of Bessarabia, he gave me a three-day furlough to rest up from my journey. Though none too enthusiastic at the prospect, I spent the three days with Irina. She was living in Arcadia, in an apartment belonging to some friends. She fed me very well; I slept or rested on a deck chair on the terrace of the villa. Arcadia in spring was very green and beautiful; the smell of the lilacs, the sight of the cherry and apple blossoms, reminded me of Cracow. The sky was blue and cloudless. The political firmament, however, was not so clear. Though there were no audible signs of thunder, you could sense trouble in the air.

Immediately after my furlough I was put on guard duty. My old story—that I had to get my truck repaired—no longer worked. Shaposhkin even took the trouble to go himself to the parking lot to find out from Soloviey whether I was telling the truth. When he came back he gave me a long lecture, during which he expressed his very definite opinion that I was the worst son of a bitch the world had ever known. That afternoon I was on my post. It wasn't a bad post. I had to stand in the guardhouse corridor and guard the prisoners. The regimental prison consisted of two large cells—one the so-called "simple jail," the other the "severe jail." There was a door to each leading from the corridor, and a third door connecting them. The guard on duty was given the keys to all three.

In the simple jail sat my old pal, Kolka, the cook, who was doing seven days for disobedience. With him were two other men. The severe jail held four. I had seen Kolka as soon as I arrived, and talked to him through the window that gave onto the parking lot. He told me that the other cooks, in some mysterious manner, had just brought him a big basket filled with sausage, bread, and vodka and that he intended to give a party for all the prisoners. To make this possible he asked me, while on duty, to open the door connecting the two cells. I agreed, of course, on his assurance that the party would be conducted in perfect silence and that it would not last long.

I took over my post at seven in the evening. A few minutes later I went to Kolka's cell, unlocked the connecting door, and returned to my place in the corridor. Through the spy-hole I observed the seven men sitting comfortably on a wooden bed. Then, from under it, Kolka produced supplies of vodka and snacks, and the party began. At eight o'clock I was to return and, after separating the prisoners, lock the door.

Well, I cannot say that that party was particularly quiet, as Kolka had promised. Several times I had to calm the revelers, though it certainly did not surprise me to see the vodka intoxicating them so completely. The drink particularly affected those confined to severe jail, who ate only one meal a day and were undoubtedly very hungry. Suddenly, from outside the building, I heard a command that made me jump.

"Guard, to arms!"

This, of course, meant the appearance of the officer on duty or someone on a round of inspection. Promptly all sentries inside the building ran out. Then I heard the sergeant of the guard shout "Present arms!" The visitor, whoever he was, had arrived. My hands trembling with fear, I opened the door to the cell and yelled: "Everyone back to his own cell! Commission!"

I was so excited I couldn't find the right key to the door between the cells. At last I found it, slammed the door, locked it, ran out into the corridor, locked that door again, then stopped and quite calmly leaned against the wall.

"Well," I sighed, "everything's all right and on time."

At that moment I heard a chorus of voices outside:

"Good morning, *tovarisch* major."

It was the commander of the regiment! He was greeting the sentries, according to regulations. I straightened out my uniform and fixed the cap on my head. The major, accompanied by the officer on duty and sergeant of the guard, entered the corridor. He took a look at the room where the sentries were resting, grimaced when he saw the blankets in great disorder, then came marching in my direction to inspect the guardhouse. As he reached me I presented arms and reported officially that I was Private Virski on post Number One, the number of prisoners doing simple jail— three, severe jail—four. The major nodded his head and ordered me to unlock the door to severe jail. I turned the key in the lock, stepped aside, and, with my back to the door, let the inspection pass. They walked into the room. Then followed a long silence. I felt anxious, without knowing why. Suddenly I heard the major's voice, directed at me:

"And where are the others?"

What others? I thought feverishly. I walked into the room, and immediately my legs began to bend under me. In the corner, on the stone floor, lay a single prisoner, snoring loudly. Without a word the major marched out into the corridor.

"Unlock the simple jail!" he snapped.

I pushed the door and peeped inside first. Oh, horror! On the bed and along the walls six men were lying. The room smelled more like a saloon than a prison cell. Behind me, the major, his nose in the air, stared at the six distinctly drunk men.

"What does this mean?" he asked sharply.

What could I say?

"I don't understand, *tovarisch* major, what's happened here," I replied weakly.

The major was furious.

"It's your duty," he snarled, "to see to it that nothing happens. Take off his belt," he added, turning to the sergeant of the guard, "and lock him up!"

I unbuttoned my belt and handed it to Jan, who was aide to the sergeant.

"Which room?" I asked, turning to the major. "The simple or the severe?"

"Severe!" he roared, "and even that's not bad enough for you!"

I walked into "severe," whose only prisoner was still snoring. He, incidentally, was one of the simple jail men. Then I heard a tremendous rumpus next door, and questions being asked in a loud voice. I began to think. Suddenly I realized that while trying feverishly to find the right key to the door, I had forgotten one thing—that the whole party was drunk. Although they had made frantic efforts to conceal traces of their debauch, they had not dispersed, as I'd expected, to their respective cells, so that the inspection found them in precisely the same positions as when I had locked them in.

The empty bottles and the remains of sausage were found under the bed. The inquiries next door were now extending all the way to the kitchen, the only possible source of the sausage. I could hear Kolka's stubborn defense. He was maintaining that I knew nothing of the whole affair, that they had opened the door with a passkey, which later, during the alarm, they had thrown out of the window. I imagined the rest of the men nodding in assent. Then I heard no more.

When night fell, I lay down on the stone floor beside the still snoring drunk, put my hands under my head, and fell asleep.

In the morning Jan smuggled me some breakfast, to which I was not entitled, since I was under "severe arrest." When the hour to report came, to my great surprise I was let out and returned to the barracks. I had been exonerated from all connection with the debauch by the unanimous defense of the prisoners!

★ ★ ★

GREAT preparations had been made for the May Day parade, which, to our surprise, took place without any war equipment—no guns, tanks, or even cavalry. Kalugin, a frequent guest in our dormitory, insisted that this absence of arms was due to military secrecy, particularly on this occasion, since the parades would be admired by German visitors.

I again managed to avoid the parade by explaining that I had left my summer uniform in storage. I was lucky, for the regiment left at four o'clock in the morning and returned at four in the afternoon. After a late dinner our commander delivered a short speech emphasizing the Soviet nation's confidence in us—the new citizens and defenders of the "Socialist Fatherland," which had allowed us the honor of wearing the uniform of "the only democratic army in the world, whose purpose is not imperialistic conquest, but the defense of all free peoples." Now, in appreciation of our progress in military training, continued the commander, the Fatherland was to confer upon us the rank of junior sergeants. So long as the battery was in training, however, we were entitled to wear the insignia of this rank only on passes, and we were to continue to receive a private's pay—the fabulous sum of ten rubles a month. Should a man be transferred from the battery elsewhere, he would automatically acquire all the rights of a noncom.

"They might just as well go the whole hog and make us colonels!" muttered Dziunek bitterly. "With a private's pay, of course!"

★ ★ ★

ONCE more at the end of May, we had to go to the polygon at Novomoskovsk. This time the departure included the entire regiment with full war equipment—all its guns, tractors, and trucks. We left in two transports, one composed of guns and tractors on open trolleys, the other of men and trucks. The journey took five days. At the stations we passed other army transports, almost always artillery, going in the same direction. The Novomoskovsk polygon was to assemble the entire artillery of the Odessa district for artillery maneuvers. At Voznyesenskaya, where the train stopped for a whole day, Kalugin came to us, looking very worried. He talked of the absurdity of sending artillery into the interior and leaving the frontiers unguarded at this moment which, to him, seemed so uncertain. Along the entire Rumanian border, which was under the command of the district of Odessa, there was apparently not one artillery regiment, light or heavy. This was indeed disquieting news. Kalugin also maintained that the Odessan staff was riddled with traitors, who were trying to divert the attention of the population from the possibility of a war which he believed would break out not later than the fall. Karpyenko, who had been observing our group with an unfavorable eye, approached us after Kalugin had left.

"Well, did Kalugin threaten you with war?" he asked with an ironical smile.

Of course not! We could not take a definite stand on Kalugin's prophecies. Though most of us agreed with what he said about the possibility of war with Germany, about false propaganda by the staff, we felt that his fears were exaggerated. After all, wasn't the Russian intelligence service aware of what was happening in the West and what could be expected of Russia's present ally? On the other hand, we had not forgotten war's sudden outbreak in September, 1939; we had experienced on our own skins the power of the Nazi "blitz" during that short but tragic struggle. We knew that diversion from within meant much more than thousands of airplanes and tanks in open battle. The German fifth column at that time had worked wonders. While large armies had been holding maneuvers hundreds of miles from the frontier on the day war broke out, whole divisions of our bombers had been destroyed in a few minutes on the airfields.

Here in Russia the fifth column could have found masses of adherents from among Ukrainian nationalists or from any nation in the Caucasus, where ideas of independence were still not forgotten. And what about the six hundred thousand citizens in the German Volga republic, or even Odessa's fifty thousand inhabitants of German origin?

Such were the thoughts Kalugin put into our heads on this journey which, incidentally, did not end without adventure.

At the station of Kirovograd, Lolek left the train with a pail to get some water for our gang. We watched him cross the tracks, which were packed with long lines of freight cars. He had passed out of sight when, without any warning, our transport began to move. Single soldiers who, like Lolek, had gone to fetch water or fruit, started running across the tracks. But Lolek could not be seen. We stood in the doorway of our freight car shouting and whistling through our fingers to alarm him. With the train increasing its speed, we passed some coaches which had stood between us and the station, and suddenly, several hundred meters away, we spotted Lolek. He was standing in a seducer's pose, his short legs crossed, leaning against a well, calmly talking to a girl rich in curves, who was pumping water. We howled and waved to attract his attention. Hearing the screams, he turned around and waved merrily at us—not having the remotest idea who we were. We roared so that we must have been heard two kilometers away. Lolek still didn't recognize us. He stopped waving and turned to the girl. We groaned with horror. An instant later he whipped around again, opened his eyes wide, and, spilling water from the pail still in his hand, began galloping toward us across the tracks. His fat, short legs were flashing at an incredible speed. Though we realized he stood little chance of catching us, we went on roaring, trying to speed him up. The pail still in his hand, he was exerting his last ounce of strength. At last, deciding that his efforts were in vain, he stopped and stood there, staring at the disappearing train like the biblical wife of Lot. Then we lost sight of him.

The situation was serious. For being absent from the transport for more than two hours, Lolek was certain to be court-martialed. We tried to think of some solution, but what explanation could we give? Fortunately none of our officers had noticed that Lolek was missing. Sergeant Dunin, the senior in our freight car, had not reported the affair; he simply pretended he knew nothing about it. We rolled on, upset and worried. Jan grieved most; it was he who had talked Lolek into going to fetch the water.

By the second day the sergeant still had not reported Lolek. The day after, when it seemed that he would have to let the *kombatr* know, a miracle occurred.

Our train had spent all morning at a small station outside Pyatikhatki. Ludwik, Dziunek, and I were lying on the grass beside the track, dozing. Suddenly we were roused by the whistle of an engine: an express train roared into the station. It slowed down a little, whistled once more, sped up, and with a great rumpus hurried off in the direction of Pyatikhatki. I was watching the departing train with indifference when suddenly Ludwik screamed, jumped to his feet, and stood staring at the disappearing coaches. I followed his gaze but saw nothing to get excited about. Then Ludwik screamed again and began waving his cap. Dziunek, quietly resting on the grass, eyed him anxiously.

"Poor boy," he concluded, "must be a touch of the sun!"

But Ludwik, fully conscious, turned a smiling face on us.

"Did you see Lolek?" he asked excitedly.

I now began to think that Dziunek was right. Although Ludwik was now quite calm, he swore he had seen Lolek in the coal tender and that Lolek had even waved his hand.

Well, it was not impossible, but none of us believed him. We considered Ludwik to have had a hallucination. After a while our train moved and we dragged on toward Pyatikhatki, coming to a halt between long lines of freight trains. We could see neither the station nor the express train, which should still have been here, according to our calculations. We didn't leave the transport, since we weren't sure how long we were going to stop. All of a sudden, from among the freight cars opposite, a queer figure appeared. He was so covered in coal dust that he was recognizable only by his short legs. Mumbling something unintelligible, he jumped toward our door. Howling with joy, we pulled him inside. Lolek was gasping with laughter. Of us all, Dunin was most pleased. A big scene would have been waiting for him had Lolek not been found. We took off Lolek's foul uniform and helped him wash. Splashing the soap in all directions, he began telling us his adventures. At Kirovograd the military station commander had wanted to arrest him. Having somehow wiggled his way out of that, he had jumped on a passing freight train and, sitting on the bumpers, had spent two days covering the hundred kilometers to Znamenka. He had no idea whether our transport was in front or behind him, nor whether he stood any chance of catching us up. At Znamenka, he had been marooned until this morning when someone had pointed out to him an express bound for Dniepropietrovsk. He explained his predicament to the engine driver, who told him to get on the coal tender and, as payment for the ride, shovel coal into the furnace. In the belief that the more coal he threw into the furnace the sooner he would reach us, Lolek shoveled furiously for hours without stopping. In fact, he had just taken his first breather when he recognized Ludwik lying on the grass. Even then he was terrified that by this time his absence had been reported and all his efforts had been in vain. Now, however, hearing that nothing terrible had happened as a result of his adventure, he was mad with joy.

★ ★ ★

THE artillery camp at Novomoskovsk looked very different in summer. The billeting area, containing thousands of tents, lay along one edge of the polygon. The entire artillery of the military district of Odessa, and most of the regiments from Dniepropietrovsk and Kharkov were to meet for maneuvers.

Our regiment's area was very pleasant. Tall pine trees shaded the tents from the burning rays of the June sun. Not far away there was a brook where one could swim. Our tents were made of such poor material that it had to rain for only an hour for the water to come through. Despite this, we felt fine. The change from the stifling barracks to the well-cooled tents was a relief. Also, the food was far better than in Odessa.

On the enormous polygon the engineers had already built many different kinds of *papier-mâché* villages and towns, to be used as targets. We fired at these from a distance of ten to fifteen kilometers. The guns were very exact, the observation and measuring instruments beyond reproach. Our battery beat the whole regiment in precision and results. Julek and Misha were the regimental champions of artillery calculations. Karpyenko was as proud as a peacock of our battery, although the credit, of course, should have gone to Kalugin.

Occasionally strange things happened on the polygon. Other regiments would fire across our range, or two or more regiments from various locations would fire at the same targets simultaneously, making it impossible to gauge the results. In the middle of June we held a firing exhibition before a special commission from Kiev. The range of our fire was to be twelve kilometers. I was on the observation post, about four kilometers from the target—a "barrack" on some sand hills. Some five hundred meters to our left, at a few tables on the edge of a wood, sat the commission and a group of high ranking officers. We on the post had dug ourselves in. With helmets and artillery binoculars, we looked and behaved as warlike as possible. We sat waiting for a radio signal to start firing. But when the regiment on our right began shooting at the target, we also let go—first, single shots, so as to observe the results, then as a battery, next as a division, finally as a whole regiment. The targets were concealed from our eyes by a dense cloud of dust and smoke. The air observer with whom we were in radio connection reported that the firing was precise and effective. Suddenly, about eight hundred meters in front of us a great plume of fire and smoke blossomed out and a terrible explosion shook the air.

Before we had recovered from the shock, another explosion, this time even closer, stupefied us. Through the microphone Jan was demanding our firing crews to tell him what the hell was happening. They answered that they were firing over the same range and couldn't understand what he was complaining about. Then a third shell burst halfway between us and the commission sitting at the tables. At that, panic broke out among the officers, some of whom began running away; others, not knowing what to do, just stood helplessly around the tables. Over the radio someone was yelling orders to cease fire at once. Our battery had been silent for some time. Then came the next explosion, two hundred meters from the tables. In hopeless disorder a mass of men took to their heels and fled toward the wood; the tables were turned upside down; the papers, blown sky-high by the force of the explosion, came fluttering down like snow. More shells exploded, one after the other, each landing nearer the wood as if in pursuit of those escaping. The next salvo passed over our observation trenches. We threw ourselves flat, our faces pressed to the ground. The cannonade must have lasted ten minutes.

"Hell," cursed Jan, lying beside me on his face, "there's war for you—and without an enemy!"

When the firing calmed down, men began hurling questions and curses at one another over the radio. In the wood there was utter confusion: one man had been killed and several wounded. Everyone was ordered to remain at his post until further notice. An investigation was started. It lasted several hours. Late that night we heard that one of the 306mm. howitzer regiments, as a result of a mistake made by its observer, had begun firing at a range three kilometers too short.

The following day we learned from Karpyenko that eight officers and twelve enlisted men had been pronounced guilty and had already left under escort to Dniepropietrovsk to be court-martialed.

"They might just as well have shot them right here," added Karpyenko.

A few days passed without exercises. We wandered about in the wood, swam in the brook, and were happy in our freedom.

On the morning of the nineteenth, we went out on maneuvers again, with full equipment. The first day and night we marched thirty kilometers outside the polygon area. Next morning we moved onto the polygon, dug our observation posts, and soon after dawn started firing. At noon it began to rain. It continued raining all that afternoon and night; all next day while we were marching and firing it never stopped raining. It rained so hard that from time to time we had to halt and empty our boots. On reaching our tents late at night, we found them under water. Exhausted, we lay down, paying no attention to the water either from above or below. That was our last night of peace.

PART TWO

THE morning of June 22 was magnificent. From the soaked tents, from the ground, and from the trees a thick mist rose into the sky. The sun was merrily warming the countryside. We dragged ourselves out of the wet blankets and hung our clothes on ropes and branches. Since it was Sunday and no duties were expected, we wandered among the tents in bathing trunks. Contrary to regulations, we had breakfast in the tents, as our uniforms were not dry enough to put on. The atmosphere was like that of a lazy holiday. The sun grew warmer and warmer.

Suddenly, from the regimental H.Q. we heard yells, commands, a general commotion. Messengers began running in all directions. Clearly something extraordinary had happened, but nobody knew what. Karpyenko burst through the tents.

"Get dressed! There's to be an assembly of the whole regiment!"

We put on our still humid uniforms. In the neighboring regiment from Kharkov, the same feverish commotion reigned. We asked Karpyenko what had happened.

"I don't know," he said. "But something very important."

"Maybe a war," suggested Walter.

We looked at him as if he'd gone crazy. Here in this wood, shut off from the rest of the world, we hadn't given a thought to the possibility of war. It had ceased to exist for us. The sun shone, the air smelled of resin. What war?

Half an hour after the commotion had started, the regiment was assembled on the square in front of the mess. In the center two trucks took the place of a tribune. On one of them stood a tremendous portrait of Stalin; on the other that of Voroshylov. The heads of the crowd moved in waves. We did not have to wait long. The major and the commissar stepped onto the tribune. The commissar raised his arm, indicating that he was about to speak. A tense silence fell over the square.

"This morning," began Shurin, "our deadly enemy, Germany, bombed our cities and villages without warning. The commissar of foreign affairs, Molotov, pronounced a historic speech declaring war!"

At first there wasn't a sound. Then the crowds began to mutter their disbelief. Suddenly, from the loud-speakers we heard Molotov's voice. His speech left no room for illusions. The bombed cities—Lwów, Kiev, Odessa, Leningrad—the dead and wounded, the Russian frontiers crossed by German armies on many sectors. Yes, this was war. Jan was standing next to me, his face very serious. I looked around. Everybody had the same expression. We were all thinking of our families, of the fact that a new era had started. War. For us this had never been an empty word. Today, as trained army men, it meant we'd wake up one of these mornings and find ourselves on the front.

The commissar was yelling something more about the Fatherland, our duty, the deadly foe (who only six hours ago had been our powerful, faithful ally). Then the major spoke along the same lines. One thing he said was interesting: turning to us Poles, he emphasized that we would be able to take revenge on the Germans for September '39. Further, that all misunderstandings between himself and us (we presumed he was referring to the prohibition of the Polish language) would immediately disappear as outdated. The band banged out the "Internationale." Everybody began to sing. In a few minutes, without command, we dispersed to our tents. Before dusk we were to liquidate the camp and march to the Novomoskovsk station.

We wrapped up the tents in silence, letting our thoughts wander far away.

"Well," said Dunin, who happened to catch my eye, "in a month we'll be in Berlin!"

I didn't answer. Against my will I found myself thinking of September 1, 1939. The people on the streets of Polish cities were making dates for a glass of beer in Berlin, also a month hence. Four weeks later the Germans were drinking beer in shattered Warsaw.

Having finished liquidating the camp, we waited for the order to march away. The atmosphere was morbid. Men sat alone, silent, on their folded tents. No one had anything to say.

Only when we started loading the trucks and a column was formed, did this gloomy atmosphere disappear. We all began talking fast, arguing as to which front would be the first in our war existence. Everybody felt optimistic and full of hope. Some men began to sing. We stopped gazing up at the sky, where, ever since the commissar's speech, we half expected to see formations of German planes. We knew, of course, that we were out of range: six hundred kilometers by air to the Rumanian border, and probably nine hundred to the German frontier in Poland.

When our turn came to move the whole regiment went on foot, the commander and the commissar at the head, the guns and trucks in the rear. As we passed through the village of Orlovaya, the children came out of their houses and, standing on the sides of the road, threw flowers at the soldiers. The women produced whatever they had—milk, bread, apples—and the men gave us tobacco. "Long live the Red Army!" and "Death to the German bandits!" they shouted. This manifestation was undoubtedly spontaneous and sincere. We were gaining courage. There was no doubt that the Germans were a terrible foe, but Russia's power was also great: the power of her human masses, the patriotic power of a nation which had been and would always be valorous, capable of fighting to the last. Russia, moreover, would not be fighting alone. In the West she had England, who would now automatically become an ally, and between them the nations of defeated Europe who would be looking hopefully to the East.

On a hill before Novomoskovsk we stopped. Beneath us we could see tremendous masses of men and guns loading on endless lines of red freight cars, then moving slowly away from the station. There was something impressive about this scene. One train after another, each composed of at least a hundred freight cars, steamed out of the station every few minutes. We realized then the vast number of men the polygon had held.

Our regiment was not to leave until the following morning. Meanwhile, we were marched away from the station to a neighboring forest, where we spent the night. It seemed idiotic to be doing guard duty in so peaceful an area and with a few machine guns against a possible air raid. There was something theatrical and unreal in these regulations, hundreds of miles from the front.

We were wakened when it was still dark and, with the aid of searchlights, loaded onto open trolleys and freight cars. Before dawn we moved slowly in the direction of Dniepropietrovsk.

As day broke, we realized that we were traveling in a very strange manner. Not more than eight hundred meters in front of our engine we could see the rear car of the transport preceding us; at the same distance behind our transport came the engine of the next. On each engine's coal tender, in the center and at the end of each train, machine guns had been placed on platforms. These comprised our antiaircraft defense. They did not inspire much confidence, but they were better than nothing. At Dniepropietrovsk we received great quantities of ammunition for small arms, machine guns, and guns. I watched Ludwik gazing at the cases of explosives arranged in pyramids on the open platforms.

"Just to make sure," said he, "that we'll be blown sky-high if we're bombed!"

From Dniepropietrovsk, all along the tracks, people waved their hands and flags at us, threw us fruit and cigarettes. Somehow each of us already felt like a hero, like a defender of women and children and annihilator of the German *Herrenvolk*. From newspapers and rumor we learned that on the Polish front the Germans had advanced by means of their surprise attack, but that on other fronts nothing had happened. Bombing on the front sectors had been very heavy, but the losses among German pilots enormous. Our Soviet planes had apparently bombed Berlin (a lie, of course); everywhere the morale was wonderful.

At all stations on this second day of war, the atmosphere was calm; the people showed great trust in Russian power and, above all, gratitude toward the army. At larger stations refreshments and meals were served free by the population or factory committees. In conversations, we kept on hearing the unvarying belief that Russia was so strong and well prepared for war that we had nothing to fear and could look with faith into the future.

"It is not for nothing," people said, "that for so many years we have been pulling in our belts, refusing ourselves a better standard of living. After all, the first purpose of the last five-year plan was to raise the army to unconquerable heights. The Germans have neither such powerful tanks nor such planes as we have. Above all, they haven't as many!"

At these moments I almost loved the Russians for their will power, their faith in their own strength, their high morale. They seemed so close to us at this time, and their position so very similar to ours in September '39. I spoke my thoughts to Jan and Ludwik.

"I admit," said Jan, frowning, "that I have grown to like them very much. As a nation they are good, sincere, and near to us. One must always distinguish between the nation and the regime. It's a nation that has a big heart, a great sense of hospitality. The Russians are people of rare kindness, but, forced to live in this iron regime which makes automatons out of them and blind executors of their chiefs' will, they are very unhappy. Unfortunately the younger generation, having no chance to compare, brought up in a manner similar to that of the Nazis, has lost these good qualities and must assimilate new ones, such as this blind loyalty to the party and a noncritical attitude to its laws."

"There's more in it than that," Ludwik broke in. We've got to admit that the Bolsheviks have raised the nation considerably. Russia under the czars was comparable to the Middle Ages; its slavery had no equal in Europe. The ignorance of the masses, the illiteracy, the clerical witchcraft, subordinated to the demands of the czarist regime, were terrible. Conditions today are far better than they were before the Revolution. Illiteracy is being fought with good results, and people are becoming a little more civilized. Slavery of masses of peasants under one landlord no longer exists. The industrialization, the development of communication, motorization, and education are all worthy of admiration. On the other hand, communism has bred a new prototype of a Russian slave, the slave of the party, of a regime without scruples. The iron police system of NKVD is a thousand times more terrible than the *Okhrana* (the czarist police). The NKVD holds in its power the life and death of every individual in Russia. Without giving reasons, it sends men and women to prisons and forced-labor camps in the most damned and godforsaken places in the world, where they spend ten to twenty years rotting to death, often not knowing the nature of the crime for which they are being punished. On the rare occasions when a man is freed after ten years of torture, he is informed briefly that his arrest was a mistake.

"If now, faced by war, the people are confident in their army and optimistically inclined, it is doubtless due to the Russians' highly developed sense of patriotism, which commands them to forget all their wrongs and remain loyal to the governing elite."

★ ★ ★

BEYOND Kirovograd, instead of turning south toward Odessa, we continued in the direction of Piervomaysk. In Piervomaysk we saw the first signs of war. As a result of a heavy night bombardment, the station was half destroyed, the industrial section of the city in ruins. The raid had caused several hundred casualties, among them a hundred children from an orphanage. The people looked calm, but obstinate, angry. Outside the city, the country was pocked with bomb craters. At Bauta the train stopped to refuel, and we got out to stretch our legs. Looking up, we could see the small puffs of shells bursting, small shining dots in the clear sky. We could hear the mournful rumble of engines, that familiar sound of two years ago. We stood there, our necks craned. Jasiu, standing near me, cursed softly.

"Sons of bitches! That's just how they flew over our country!"

Suddenly a shower of light bombs crashed on the station, on the tracks. The noise was continuous; dust covered everything. Grabbing Jasiu by the sleeve, I jumped into a bomb shelter. The hell of destruction lasted a couple of minutes, then calmed down. Jasiu and I emerged from the shelter. The station was on fire; so were the front freight cars in our transport. Those in the rear were destroyed completely. The transport ahead of us was almost totally wrecked. Our battery, however, had not been touched. None of the boys was even wounded. We rushed to help those in the front freight cars. Regardless of the flames and explosions, everyone joined in throwing out the cases of ammunition, evacuating the wounded and localizing the fire. In an hour the last flames had subsided. A new engine was attached to our transport and we moved off, leaving behind thirty dead and nearly a hundred wounded. This first bombardment made a deep impression on our Russian colleagues, not because they lacked courage, but because they had never experienced anything like it before. We, who had watched German planes flying over Polish rooftops at will, firing at pedestrians in the streets, knew what to expect from air raids.

After leaving Bauta, we had one raid after another, fortunately none so severe as the first. As a result, the train slowed down, stopping at every alarm, while we dispersed into the fields on the order of H.Q. Only the machine-gun crews remained on the train—utterly useless, of course, since machine guns could do no damage to high-flying bombers.

During the night of June 24 we reached Tiraspol, our destination. The station, completely blacked out, was packed with military transports, trucks, automobiles, horses, men, and all kinds of military equipment. We began the gloomy business of unloading in the dark. Before dawn, Dunin sent Jan, Julek, and myself up a high wooden observation tower to man a machine gun. We stood staring into the darkness, straining our ears for the moan of German planes.

As it grew lighter, we could see the men below us feverishly unloading the transport while others moved this mass of material away from the station. Suddenly, high above us, a machine gun barked, showering the station with tracer bullets. Simultaneously we heard the moaning of engines. I stared skyward, holding the trigger of the heavy machine gun. Another machine gun started firing, sending burst after burst at the invisible enemy. Suddenly I saw something dark, like a shadow, crossing the gray sky. I took careful aim with my sights, pulled the trigger. The gun belched a long fusillade from all four barrels. I felt a relief in my strained nerves. I knew I had no chance of harming planes flying at three thousand meters, but this firing gave me the feeling that we were not defenseless. Beneath us there was pandemonium. We could see men rushing off into the fields, others still trying to unload the trains. Bombs were falling into a small forest near the station. But the next ones hit the tracks, the trucks, and the men. The sky was now almost clear. Our machine gun spat burst after burst. Others near us were also firing. Julek was drawing Jan's attention to something under our tower. Before loading the gun again, I looked down. Immediately below us lay a heap of small aircraft bombs, no doubt unloaded from the train during the night. I shuddered. Jan replaced me at the trigger.

For a short interval the bombing stopped. Suddenly, from above us, came a whining noise. Quickly it grew into a scream. We knew this scream—the planes were diving. Jan was firing like mad. Several bombs crashed together. Our tower trembled, but held up. The power of the blast knocked all three of us flat on the platform; the gun, bent over sideways, was leaning against the wooden rail. Julek jumped up first and grabbed the gun's handles. Everything was still blacked out by smoke when Julek began to fire again. Underneath us terrible things were happening. A thick wall of smoke was rising to our level, with columns of fire and explosions bursting out of it. The planes, dropping to a height of four hundred meters, were dive-bombing the station, riddling it with their machine guns. All our guns kept firing like mad. One plane, heading straight for the station, suddenly burst into flames and crashed to the ground. Though probably not touched by our fire, we yelled in triumph. As the smoke below us began to lift, we looked down on the burning freight cars, trucks, on the dead and the wounded lying in puddles of their own blood.

But we didn't look at this sight very long; the planes were returning. Our gun shook again in a fury of firing. All of a sudden Jan screamed. I thought he'd been wounded, but no, he was pointing downward. I followed his gaze. The heap of bombs below us was surrounded by a thick wall of smoke and from behind this wall tongues of flame and showers of fragments kept shooting out. The base of our tower had caught fire. We could feel the heat rising toward us. By now the planes had flown off. Above the forest we saw another one crash, dragging behind it a long column of fire. We stopped shooting and looked below.

What had been the station was crowded with soldiers returning from the fields and shelters. Now began the saving of the wounded, their evacuation to areas away from the station, the extinguishing of the fires. But we were marooned. Nobody could approach within 150 meters of our tower. The bombs below us were bursting with clock-like regularity. The tower, made of pine boards, was quietly, evenly burning; the fire had risen halfway up.

"Well," observed Julek with a certain melancholy, "the birds are going to be roasted on their branch."

Our situation seemed perfectly hopeless. If we attempted to go down we'd have been promptly torn apart by fragments of shells, and our remains burned to a cinder by the rising flames. The smoke was getting into our eyes and throats; we began to cough. Through the smoke we could see Karpyenko running around in a circle, pointing at us, yelling, then turning back in his tracks. I gazed around, seeking some means of escape. Then I noticed, some eighty meters away and at the same height as our platform, part of the destroyed station sticking out toward us. It was a section of wall, with a window more or less intact.

"If they could throw us a rope from that window," I said without conviction, "maybe we could get across."

Julek and Jan jumped at the idea. We began yelling at the men below, pointing at the station ruins. At last they seemed to understand. But how were they going to get into the ruins? The road to it was still hot with burning wreckage. Then, behind another building, we saw a group of men.

We recognized its leader, Karpyenko, then Jasiu, Lolek, and, of course, my Aram. After the explosion of a bomb they took advantage of the few seconds before the flames would reach the next one, and dashed out toward the ruins. They had barely made it when the next bomb exploded. No one fell. They all jumped behind the wall. A minute later we saw them in the window at our level. We stared incomprehensively at their mysterious preparations; none of us could see how it would be possible to throw a coil of rope from that distance and hit our platform.

After a while we began to understand. Karpyenko and Aram had fixed to the window one of those small catapults which can throw a telephone cable a distance of 150 meters. We had often seen it in action during signal corps maneuvers. The first two shots were not well aimed; the coil of rope flew wide and dropped far from us on the ground. The third, however, was masterly; the coil whistled over our heads, its weighted end flying beyond us. Julek and I grabbed the cable, which tried to escape from our hands. It tore the skin from our palms, but we held on. We could hear Aram howling with joy in the window. In our excitement we forgot the bombs and even the flames, though they were now no more than three meters from our platform. With bleeding hands we hauled in the cable, to the end of which was attached a length of thick rope. This we tied to a wooden beam of the tower. While we were stretching the rope, we heard an ominous crackling. Looking down, we saw that part of the tower's base had burned through and carbonized.

"Well, if this damned gallows doesn't collapse," mumbled Julek, "it'll be a miracle."

We decided that Jan should be the first to attempt the crossing, as he was stoutest. I was to be last. With no confidence in his ability as an acrobat, Jan looked at us sadly, hanged himself up by the hands and feet, and began his crawling. Halfway across, as the rope sagged under his weight, he slipped. We thought he was done for, but he managed to recover, exerted the remainder of his strength, and slowly, hand over hand, reached the window, where Karpyenko and Aram grabbed him in their outstretched arms.

At the same moment Julek threw his legs over the ramp, grabbed the rope, and skillfully, like a monkey, began to make his way across. When he had reached the middle, the fire suddenly began to attack the platform, fortunately not from the side where the rope was attached. Seeing this,

Karpyenko yelled that I should get onto the rope without waiting for Julek to reach the end. I grasped the rope, threw my legs over the rail. Because Julek was two-thirds of the way across, the rope didn't sag at my end, so that I couldn't slip down it the way the others had. Crawling along that rope was much more difficult than I'd expected. My torn hands hurt like mad. Under me I heard the whistles of bomb fragments, now bursting in salvos. Julek, whom I could just see out of the corner of my eye, had reached safety. I needed every ounce of my strength to drag myself to the end. The moment I felt somebody's strong arms go around me I lost consciousness, but only for a fraction of a second. The instant Aram and Lolek laid me out on the floor and poured water over me, I was all right. I got up after a while and went to the window. Julek, Jan, and I stood looking at the tower which had so nearly been our grave. The rope had burned through; so had the platform. Then the wooden construction, the platform, the four-barreled machine gun, came tumbling down with a crash.

Later that morning the regiment, having lost thirty men in the bombardment, quickly moved to the forest, a few kilometers from Tiraspol. We stopped near the river Dniester on which a second defense line had been prepared. For the time being the front remained on the river Prut, at this point some eighty kilometers from the Dniester. The raid had had a bad effect on the men; their morale had been shaken.

After dinner we saw formations of German planes flying quite low. Apart from ack-ack, no counteraction was taken. We were amazed. Where were our fighter planes? One German plane must have spotted our group in the forest, for it dropped out and sprayed us with a rain of machine-gun fire. Fortunately no one was harmed.

This first night at the front proved a nightmare. Near us was a road on which supply convoys were continually passing. All night long German planes kept roaring over the road, firing blindly from machine guns. From time to time they dropped parachute flares. These flares were able to remain in the air from two to three minutes, while the tremendous area was bathed in light. Next morning the regiment was ordered to Byeltze, on the Prut. Many of the regiment's drivers, including myself, were assigned to divisional H.Q. to transport the ammunition dump from Tiraspol to the front line. After this job I was to return to the regiment.

Cursing my damned profession, which for the second time was separating me from my comrades, I packed up and drove my truck to the division, a few kilometers from Tiraspol. On Karpyenko's orders I had attached to my collar one triangle, my new rank of junior sergeant. At H.Q. they were already organizing a number of convoys, each consisting of forty or fifty trucks. A junior *politruk* with a gay face eyed me carefully.

"Sergeant," he asked, "how long have you been a driver?"

I wanted to throw something at him, but seeing his pleasant, open muzzle, I told him four years. Smiling with contentment, he ordered me to stand aside. He then organized a group of forty-five drivers, made me head of the group, and presented himself as the commanding officer of our convoy.

Our job was to deliver artillery ammunition of all calibers from Tiraspol to different points on the front line; also ammunition for rifles and machine guns for the infantry on the river. We were like moving gunpowder factories, traveling under a sky filled with the sound of enemy planes. Closer to the front line, we were fired at not only from the air but by artillery from across the river, the shrapnel bursting in clouds above us and covering the ground with fragments. Down by the river, bursts from heavy German machine guns rained on the trucks; sometimes even antitank projectiles came whistling over from the other side of the river. These rides are something the survivors are not likely to forget. The H.Q. was not concerned with losses; no matter what the cost, this mass of ammunition had to be transported from the supply dump, where it lay exposed to total destruction, to the formations which so desperately needed it.

After the first day of shock I grew accustomed to the thought of death, to the idea of being hurled into the air with a load of explosives, to being killed by the fragment of a grenade or a machine-gun bullet fired from the air. At first bewildered, I slowly became indifferent to my fate. Without sleep, I fell into a state of chronic lethargy. My *politruk*, good man, who always rode with me, kept me going on coffee and bread and bacon which he thrust down my throat almost by force. Day and night we didn't get one moment to relax. Every morning at dawn we set out in a group of forty to fifty trucks, returning twenty-three hours later in a group of thirty to thirty-five. During that time the missing trucks had either burned, exploded in the air, or been abandoned on the roadside, total wrecks. The hour before starting out again was spent reorganizing the convoy and assigning new men to replace the losses. Every day between fifteen and twenty men were killed and wounded. Whereas the atmosphere on the front line was calm, almost idyllic, interrupted only by short spasms of firing from one or the other side, the supply convoys were continuously under attack. Seldom, not more than once a day, did we see our fighter planes intercept the enemy. Most of the time the Germans flew at will over our territory, suffering losses only from antiaircraft artillery.

After the first three days I had grown so indifferent that I did not react to exterior impressions. Like an automaton, I drove my truck, complying with the *politruk's* orders. During air raids I did not even attempt to hide; I usually stepped on the gas, shot off the road into the fields, and went on driving no matter what the ground was like. All around me I saw trucks in flames or exploding in the air, and all I felt in my apathy was a sensation of surprise that I was still well and untouched. On one occasion the rear of my truck was hit several times by a diving Stuka. The truck shuddered all over. I went on, expecting it to burst into flames or explode. The *politruk* crouched on his seat. But nothing happened. Next time we unloaded, he made me look at the truck's rear. Two inches above the level of the cases of artillery shells, it was perforated like a sieve. If the shots had hit that much lower we'd now be flying among the angels. From then on fatalism began to form in me; I became its strong believer.

On the sixth day of this nightmare the job of transporting ammunition supplies came to an end. Almost unconscious from fatigue, eyes swollen, unshaved for a week, I reported to my regimental H.Q. Although his formation was stationed elsewhere, the *politruk* accompanied me. The good man, of whom I had grown very fond, was afraid that if I went alone I'd fall asleep over the wheel and land up in the ditch. During the last couple of days his job had consisted of pulling my elbow whenever he saw my eyes were closing of their own accord, and of feeding me vodka and coffee. When he entered the hut belonging to my regimental H.Q., I sat outside on the step of the truck. A few minutes later he reappeared, accompanied by Lermontov. I had no strength to rise and salute the captain.

"The *tovarisch politruk* has reported on your courage and self-control," he said, smiling. "I am sending in the suggestion that you be decorated with the 'For Valor' medal."

I raised my head indifferently.

"I was not courageous," said I. "Just too tired to care, *tovarisch* captain."

Lermontov smiled again.

"Well, well, don't be so modest. Now you can have three days of complete rest. Nobody has the right to move you, under any circumstances. This is a provision in the order of the day. It also applies to the other men who returned. Report it to your battery."

He turned and went back to the hut. The *politruk* and I said good-by to each other very cordially.

"We've spent some hot days together," mumbled the good man. "Too bad we have to part. We were doing well."

The mechanics took my truck for a checkup. I dragged myself, barely able to keep awake, to my battery's firing posts. The guns were parked some four kilometers from Prut. There, all seemed relatively calm. I did not even look for my comrades. I found the sergeant major's hut. Shaposhkin stared at me, his eyes wide, as I pulled myself through the hole of the hut's entrance. Inside hung a smoky little night lamp. I mumbled something about the captain having ordered me three days of rest. Shaposhkin led me affectionately to a wooden bed and told me to go to sleep. He brought me some food and drink, but I was too exhausted even to look at it. Judging by Shaposhkin's almost embarrassing care, I must have been a terrible sight. Just as I was, with the belt, gun, map case, and boots, I let myself fall on the bed, covered my head with a blanket, and fell instantly asleep.

I don't know how long I had slept when a light falling straight on my eyes woke me. Blindly I reached out for the blanket, pulled it over my head. But immediately I felt somebody pull the blanket away and in a fog I heard a strange voice shouting:

"Get up, *tovarisch* sergeant!"

I opened one eye. An officer I had never seen was standing there. Bending over me, he repeated:

"Get up!"

The light from the night lamp shone on his collar, where I saw the insignia of a senior *politruk*. Even though his collar-facings were black and red, he was certainly not from our regiment. I did not even feel like opening my mouth; without trying to find out what he could want from me, I closed my eyes and again pulled the blanket over my head. But again it was pulled away, and the voice shouted:

"What's this? Get up, I tell you!"

I was furious, but I hadn't the strength to show it.

"What's the matter?" I asked weakly,

"Are you a driver?" asked the stranger.

I nodded my head.

"Then get up. We are to go to Tiraspol," ordered the *politruk* violently.

I summoned the rest of my strength to explain the situation. "*Tovarisch politruk* I have the honor to report that I have just returned from six days of tiring driving, and I have been given three days rest by the chief of staff, Captain Lermontov. I'm too weak to drive now. Take another driver!"

Having finished my explanation, I again reached for the blanket. With fury the officer pulled it out of my hand.

"I order you to get up!" he roared. "We are going to Tiraspol." I was so weak I could barely speak.

"I'm not going anywhere. I've explained why, haven't I?"

"You refuse to obey an order?" howled the *politruk*. "Do you know what that means?"

I knew. He had the right to shoot me, but I didn't care; I couldn't even think about it. I just wanted to sleep, to sleep, and once more to sleep.

"Leave me alone and go to hell!" I said feebly.

The *politruk* gaped at me, his eyes enormously wide. He stood like that, motionless for a while. Then, with a quick movement, he snatched his *nagan* from its holster, pointed it at my head. His voice was trembling with madness.

"I'm going to kill you like a dog!" he howled.

I felt no fear at the sight of the revolver, nor on hearing his threat. I was indifferent to everything. I wanted to sleep. I don't know what part of my nervous system made me do what I did next. Calmly I raised myself on my elbow and, keeping an eye on the *politruk* I quietly reached for my belt, opened the holster, drew out my *nagan*, placed it on my knee, and aimed it at the man before me. I don't know what passed through my mind at that moment. I do know that although I expected a shot and inevitable death, I nevertheless felt pretty sure that the man would not shoot.

The *politruk* stood in front of me with round, surprised, frightened eyes, without appearing to understand the situation. For what seemed an eternity we remained in this position, our guns aimed at each other. I was still waiting for the shot. I knew that if he did shoot, I didn't stand a chance: he could not miss, and I could not jump to the side or change my position. But I also knew that the moment he fired I'd have time to pull my trigger. And I knew that I would not miss.

I felt neither anger nor desire to take revenge, but one idea did come to me: I was not going to give my life away for nothing. Though I was aiming at the *politruk*, I felt my eyes slowly closing on me. My only thought was, Oh how I long to sleep! Then the minutes of immobility ended. The man, staring at me like a hypnotizer, slowly lowered his gun. I did not know what he intended to do. Slowly, without taking his eyes from my face, he put his weapon back in its holster. Finally he said:

"*Sukinsyn!*—Son of a bitch!"

But he still stared at me, while I propped myself up on my elbow and with all that was left of my energy. The revolver weighed heavily in my hand. Then, hesitating, he turned his back to me, stopped for a minute, and head bent, very slowly moved toward the hole of an entrance. Seeing his arms pressed stiffly to his sides, his head squeezed between his shoulders, I felt that he now expected a shot in the back. The mere idea made me smile. I did not want to shoot, or kill. No. All I wanted was to sleep. But that he could not know.

At the entrance he stopped again, turned around violently, and called in a voice breaking with hysteria:

"Why don't you shoot, you son of a bitch? You're going to the wall anyway!"

I still managed to keep my eyes on him. But oh, how they tried to close! Then with one leap, he was out. My *nagan* seemed to return to its holster of its own accord. I pulled the blanket over my head. This time it stayed.

When I woke, the little hut was packed with people, all talking at once. Karpyenko, Lermontov, Kalugin, Shaposhkin, Ludwik—they were all there. Because I was still terribly sleepy I presumed that little time had elapsed since my scene with the *politruk*, Lermontov, observing that I was awake, threw himself at me, yelling:

"Did you go mad, man? Do you imagine that anyone can now prevent you from getting a bullet in the head?"

Karpyenko and Kalugin were voicing the same sentiments. Shaposhkin was standing in a corner, staring at me, swallowing tears. Ludwik, pacing up and down the hut, was mumbling to himself. Everybody looked at me as at a living corpse. Nobody doubted the outcome of this affair. I had no illusions either. But all I wanted was to get my sleep. I looked at Kalugin beggingly. He came up to me and asked what I wanted.

"Sleep," I said. "Nothing but sleep."

Kalugin let his eyes rove over the men in the hut. Their faces had nothing nice to say about my mental state. Only Lermontov seemed to understand.

"He has not slept for six days!" he explained.

Without a word, they began to leave. The last of them had probably not crossed the threshold before I was once more asleep.

Next time I opened my eyes, Shaposhkin was sitting in the corner. I slowly realized where I was and what had happened. I was madly hungry. Letting my legs hang from the bed, I stretched myself out.

"Well," asked Shaposhkin with a smile, "did you get your sleep?"

Yawning from ear to ear, I nodded.

"You've slept for forty hours without once waking," he said.

I was amazed.

"And what about my trouble?" I carefully inquired. "When am I going to be court-martialed?"

Shaposhkin smiled broadly.

"Well, you've been lucky this time. Somehow the captain defended you. You have lost your medal, but kept your head."

I could not believe my ears. I later discovered that Lermontov had made a big rumpus, explaining that after six days without sleep I could not be held responsible for my actions. Without even examining me, the regimental physician produced a certificate to that effect. Lermontov had supported his defense with my "courage" and his suggestion for a medal. And from the Bessarabian H.Q. came an order freeing me of all responsibility for my "crime" of not complying with the command and of threatening an officer with a weapon. This case was without precedent, quite unbelievable.

I ate an enormous meal, emerged from the hut, and found Ludwik, Lolek, and Jasiu behind the guns. Jan, Julek, and the rest were on the front line as observers. I was greeted as someone just risen from the dead.

"We had already placed a cross on you," said Ludwik, moistening my face with cordial and affectionate kisses.

"But you will not get the medal," grieved Lolek. We all laughed. Only now did I realize the threat that had been hanging over me, the enormity of the folly I had committed.

"After all, what Lermontov stated in my defense is true," I concluded. "Had I been fully conscious I'd never have dared to do what I did."

Lolek agreed.

"Or at least you'd have done away with that man," he added, "so he'd not be able to accuse you!"

★ ★ ★

FOUR days after my great sleep Karpyenko called me. Something to do with the politruk I thought. But no, the lieutenant began asking me about my knowledge of the German language, whether I knew how to swim, and whether I was any good at finding my way about in the dark. My German, said I, was fluent, swimming the best cultivated of my sports, and owing to some eye defect I could see like a cat at night. Karpyenko was satisfied.

"Now you'll have an opportunity to rehabilitate yourself," he began mysteriously. "Our division is to send out patrols across the river—night patrols, naturally. I have chosen you as patrol commander."

I cursed inwardly. They are intent on making me into a hero, I thought to myself. I don't like the idea. I prefer to remain as I am. Outwardly, however, I had to make a pleasant face and pretend that I was very happy to have such an "honor" bestowed upon me. After dinner they called me a second time. In the room there were already twenty people chosen from the division. The idea was to cross the river upstream from the observers' posts and the infantry, and to penetrate inside the Rumanian-German positions, if possible as far as their first artillery line. Any information was badly needed. Listening to this, I began to shiver. Then Karpyenko turned to me:

"*Tovarisch* sergeant, how many men do you want to take along on this first trip?"

I answered this question by asking him another. Had I been chosen as a patrol commander because I knew German? Was I supposed to ask the first German I met on the other side for information?

"Oh, no!" replied Karpyenko. "The idea is that it's better to know the language in case you're caught."

The mere idea dried my throat up. As for Karpyenko's reasoning, I found it puzzling.

"In the event of my falling between German paws," I asked, "should I politely explain to them in German that it was a mistake on my part, apologize, and leave?"

Karpyenko did not say anything for a while. Then he collected his thoughts.

"It is always better to know the language," said he.

Suddenly in the doorway we heard a commotion. Turning, I saw Aram squeezing himself through a group of soldiers. He saluted Karpyenko, looked at me, and asked:

"Are you going on the patrol?"

I nodded my head. Aram turned to Karpyenko.

"*Tovarisch* lieutenant, I will join him as a volunteer."

The last time I had seen Aram was on the station at Tiraspol. Since then he had been transporting supplies from Tiraspol to the rear—a far safer job than mine. He had returned the day before, and, having learned of my new job, had come running immediately. Karpyenko, aware of the friendship between us, agreed without a word of protest. Encouraged by Aram's arrival, but still sad, and, frankly speaking, terribly frightened by what was awaiting me, I decided that on the first patrol we'd need only five men. I picked the men at random, without thinking. At dusk we were to go up to the infantry trenches and attempt to cross the river whenever I thought it advisable.

We were sitting in the infantry trenches trying to see something on the opposite bank—in vain, of course. From an infantry officer I had received all the information concerning the distribution of forces on the other side. Rumor had it that the opposite bank was manned by Rumanians, with a few Germans to back them up—the Germans presumably not having too much confidence in their "gypsy" allies. The bank itself was apparently not fortified, the first trenches being some three hundred meters from shore, inside the territory. While the bank on our side was flat and stony, the riverbed falling gently away, the other side was high, precipitous, and covered with bushes of osier. Right under the precipice the river was mostly over six feet deep, its current swift and violent. According to my informer, who had twice crossed to the other side, the main difficulty lay in climbing the far bank, and, on returning, letting oneself into the water without making a noise. Aram, who never left my side, was quietly cursing in Armenian and demanded to know the purpose of this adventure. I could not answer him.

At about eleven o'clock, when complete silence reigned on both sides, I decided it was time to go. The infantry officer, his platoon, and three machine guns were to keep watch until our return, and, if necessary, cover our retreat with fire. One after the other, five in all, we jumped out of the trench and crept along the stony beach. All we could see in the darkness was the vague outline of the river. I had a queer feeling of anxiety; from our very first move I expected shots from the other side. So far, not a sound. We were equally silent, for we were wearing sneakers with cord soles, in which we moved like shadows. All the buckles and other metal parts which could possibly make a noise had been removed from our uniforms. With the exception of Aram, we each had an automatic pistol hung around our necks. Aram had no use for such things; instead, in his belt he carried two long knives which he had procured in Tiraspol. The moment we reached the roaring water my anxiety left me. My mind was so occupied with calculating how best to avoid giving ourselves away by noise or movement that I had no time for other thoughts. The bank was silent. Aram and I dropped into the water. The others lay flat on the rocks; they were to wait until we had reached the other side. The water was icy. With hunched shoulders we moved on, Aram holding me firmly by the hand. The slow walk through this icy bath was extremely unpleasant. My teeth began to chatter from cold and excitement. The water now rose to our armpits. We resisted the current with difficulty. The shore was only five meters away. Coming to a silent understanding, we dipped our necks in the water, and, with a few strong movements of my legs and arms, I reached the opposite bank. I grabbed some flexible osier branches which were hanging over the water. But there was no bottom under my feet. Carried by the current, Aram landed on top of me. I held him up with my right hand while he also grabbed some branches. I thought we had made a terrific noise, but this was only the result of being oversensitive to acoustics. We rested for a minute. Aram bent over me and against the roar of water whispered in my ear that I should stay where I was while he climbed the bank. From there he would give me a sign. I watched him pull himself up, holding on to the branches, looking for an easy ascent. At last he began to climb. I let the current carry me nearer to where he was. The shore was flat here, but to get out presented great difficulties, for there was nothing to stand on. Then Aram leaned down, took me by the hand, and, with a violent tug, pulled me into the bushes. When I realized that we were now on German territory, separated from our men by a rushing river, I was again afraid. We carefully stuck our heads out of the bushes. Our eyes, accustomed to darkness, could distinguish the contours of the territory. So far as we could see the terrain seemed completely empty. Aram, squeezing my elbow, pointed along the shore. Some eighty meters above us I saw an entrenched post, probably a machine-gun nest, with sandbags covering one side of the rampart. Then I saw a human silhouette bending over, and the glow of a cigarette. We were not afraid of this post because, located so near the river, the men could hear nothing above the roar of the water. The question was: where was the next post stationed?

We began to crawl along the shore, taking advantage of the bushes for cover. Soon we noticed a similar post some eighty meters ahead. The posts were separated, therefore, by about two hundred meters. By chance our landing point had been well chosen. We returned to that spot, re-entered the water. Aram could imitate superbly the voice of an owl, of which there were many along the river. He now made two such calls. Almost immediately we heard a similar sound from the other side, from Aram's compatriot, Argebeyan, whom I had picked for patrol service on Aram's advice. He was in command of the remaining two men who had stayed on the other side. Now, on Aram's call, they were to try to join us. We remained in the water, observing the three shadows moving slowly across the river. Aram called once more. There was a slight commotion, then Argebeyan, swept down by the current, crashed straight into us. With our help, he skillfully pulled himself up into the bushes. A moment later the other two, Smirnov and Kartchevskij, were with us. Half of our job was over, accomplished with exceptional efficiency. Now for the other half!

I untied the pistol from my neck, unfolded the greasy rags in which the trigger had been wrapped. Leaving Argebeyan on our "landing abutment," the four of us began to crawl forward on a line perpendicular with the river, Aram in the lead. He moved like a cat, completely soundless, pulling himself along twenty meters and then stopping to wave us on. We had covered some three hundred meters like this without finding a trace of human existence, when, to our left, we heard some voices and saw the outline of an infantry trench. There did not seem to be any fortification either ahead or to our right. We were lying flat on the ground. Then Aram, crawling like a snake on his belly, disappeared into the darkness. After twenty minutes he reappeared as soundlessly as he had gone. His mouth close to my ear, he whispered that we should turn to our right and go straight ahead for at least one kilometer. Since the only thing I could do was to rely on Aram's sense and bump of locality, we set off. We had covered two kilometers without difficulty when we saw some light artillery posts about two hundred meters ahead. They were not entrenched, but protected from air observation by a small clump of trees. I presumed these were the guns from which we had suffered so much on the front line and which could not be identified from aerial photographs of enemy territory. The crews seemed to be asleep. We could hear only the strides of the sentries. Aram disappeared again. We hugged the ground, holding our breath, the enemy seemed so close. After a few minutes Aram was once more at my side. Impossible to proceed, he said. We turned around and began to retreat the same way we'd come. With great effort, we reached the river. Aram called again. Argebeyan owled his answer. Led by this call, we slipped into the bushes. Argebeyan reported that everything was all right The crew of the machine-gun post had been changed. According to the Armenian's observations they must have been Germans, not Rumanians.

We slipped into the river. The current carried us away with such power that we didn't dare go too deep into the bushes on our side for fear that the infantry there, not having been warned of our expedition, would take us for the enemy and start firing. For this reason we crossed many meters up the river so as to land in the same spot from which we had set out. Actually this was unnecessary, since the infantry officer, who had twice crossed the river, had foreseen precisely what we would do and had informed our posts accordingly.

It was almost three o'clock when, soaked and freezing, we were back again in the hut of the divisional H.Q. Karpyenko, who'd been asleep, leapt up on hearing that we'd returned. The staff sergeant put a bottle of vodka, a loaf of bread, and a piece of lard in front of us. We all drank half a glass, letting the heat spread through our frozen limbs. Then I sat down and gave Karpyenko a detailed report of our findings, while Aram described what he had seen from even closer range. According to him, the enemy artillery posts were placed symmetrically every fifty meters. Delighted, Karpyenko told us to go and rest. Outside the hut Aram suddenly turned on his heel and jumped back into it again. He came out clutching the bread and lard and the half-empty bottle of vodka. He grinned at us with his white wolf's teeth.

"What?" he said indignantly. "Leave all this to the staff gang?"

★ ★ ★

AFTER a week of comparatively quiet life, Aram and I were again called to divisional H.Q. During the past couple of days patrols from other divisions had not been too lucky. Several members of the last two had been killed while attempting to cross the river. According to Karpyenko, this particular night was expected to be unusually dark, and H.Q. wanted our patrol to penetrate as deeply as possible so as to identify a concentration of armored units which had been observed by our aircraft. A number of patrols were to be sent out over the entire area, each with the task of infiltrating a different sector. Having picked ten men, Karpyenko issued his instructions, basing them on the experience of those who had survived previous sorties.

"If he knows so much about it all," muttered Aram, "why doesn't he go himself!"

From the start, while lying on the ground waiting to cross the river, I had a premonition that this time things were not going to turn out so well for us. At one moment we heard a sharp cannonade from machine guns and saw columns of searchlights passing over the water. These were the first searchlights we had seen on this sector. We waited a few minutes longer and at about 11:30 began to cross, one after the other. It was exceptionally easy. In an hour we were all on the other side. I left three men on the bank, and in a group of seven we began to penetrate inside the territory. As usual, Aram crawled ahead of us—the ears and eyes of the whole patrol. We had covered at least two kilometers without observing any tanks or even traces of them. At two o'clock I decided to retreat. Having inspected the sector assigned to us, we had every right to do so. We had covered half the return trip without interference when all of a sudden Aram leapt back to my side. No more than twenty-five meters in front of us stood a German post, with two men. We could turn neither right nor left, since we knew there were posts in both directions. This post had not been there before. The situation was certainly dangerous. Our only hope was to dispose of the sentries. But how, without causing an uproar? I asked Aram's opinion. In answer, he put a finger to his lips and crawled forward without a word. Having waited fifteen minutes, I was about to start crawling after him, when suddenly, beside me, I heard a sound in the grass and up jumped the dark shadow of Aram.

"We can go on," he muttered, and held some dark object under my nose. "This is for you," he whispered. "You hate the Germans so."

Switching the revolver to my left hand, I put my right on the dark thing he was holding. It was round, warm, and wet. I couldn't make out what it was. Then I felt myself trembling. Between my fingers I distinctly felt human hair. I felt sick—it was a human head, evenly cut off at the neck. Hot blood was still running from it. Only now could I see in the darkness a pair of wide-open eyes numbly staring into the distance. Nauseated, I threw this horror into the grass.

We again began crawling toward the river. We had almost reached it when a searchlight unexpectedly flashed. I saw it catch the bent silhouette of one of my men, who was so startled that he just stood there motionless, as if turned to stone. A split second later a machine gun barked. My man fell. Simultaneously another searchlight flashed in our rear, the bright arc moving in our direction. The moment it was about to fall on us, Aram, no longer able to bear the suspense, began firing his revolver at the searchlight. The light discovered us as we lay pressed to the ground. I jumped to my feet, yelled at the men, and we fled, zigzag, to the river. Three machine guns promptly opened up on us. On my right I saw someone fall. I heard a groan behind me. Aram was stumbling. Always more at home in the dark, he couldn't keep his balance in this bright light. I clutched his arm, pulled him along behind me. Though the searchlight was following us, it had stopped for the moment on one of the men in the rear. In front of me I saw a clump of tall grass. I pulled Aram by the arm and we both fell to the ground. The light passed over us. We promptly jumped to our feet and continued running toward the river.

Our shore, wakened by the shooting, had now begun returning the fire. While running I could see the flash of red sparks on the other side. A fine thing it would be, I thought, if our own men were to do away with us!

Suddenly Aram pulled me to one side, and the next thing I knew we were both out of our depth in icy water. Beside me, Aram, breathing heavily, was grasping the bank with one hand and holding me up with the other. Above us the firing continued, while the searchlights were playing over the water. I saw one of their rays stop still on two heads which, fighting the current, were trying to cross the river. In a flash I recognized Argebeyan, who had remained on the shore with the two others. The machine gun above us immediately opened up with a fusillade, and the heads passed swiftly down the river, emerging and dipping, until finally they didn't emerge any more.

Aram began crawling along the bank. In a minute he was back again, to tell me in a whisper that the machine gun above us was no more than eight meters over our heads, manned by three men. At the moment, he said, they were busy digging in, as if they intended to stay. The situation was thoroughly bad.

We didn't stand a chance of swimming the river; we'd be spotted at once and shot like ducks. Now that the firing had calmed down one splash from us would have meant instant death. To make matters worse, when I'd fallen into the water I had let my pistol slip out of my hands. Aram still had his (I had ordered him to carry one on this trip), but by now the water would certainly have made it useless.

Our arms were growing mighty tired from hanging onto branches in a powerful stream constantly pulling us with it. I felt mine getting stiffer every minute; I no longer had any feeling in my legs; my hands and face were bleeding from cuts caused by dashing blindly through the bushes. To wash away the blood which kept streaming into my right eye, I immersed my head in the icy water. Every time I did this I had a strange feeling of laziness which made me want to leave my head under water. After an hour or so the blood stopped running from my forehead.

Despite a much stronger constitution, Aram's condition was as weak as mine. I have no idea how long we had been hanging onto those bushes when we heard a splash behind us and a man, carried by a wave, Was thrown straight into us. Instantly Aram regained all his energy. Holding onto the branches with his left hand, he grasped the man by the throat with his right. A brief rattling sound came out of the mouth of the choking man. I heard someone shout something incomprehensible, then, above us, a harsh voice in German;

"*Wer da?*" Who's there?

We held our breath. The man Aram and I were holding onto was stiff and inert. The shouts of "*Wer da?*" were repeated above us. Then they stopped.

At last it grew light enough for us to see the face of our "guest." We both gasped. It was Smirnov, the third man of the group we'd left on the bank with Argebeyan. Had Aram strangled him, I asked myself? No, not quite. Holding him by the neck, Aram kept dipping Smirnov's head in the water.

Presently he opened his eyes, then his mouth—and was obviously about to let out a scream when I clamped my hand over his jaw. When I saw that he recognized us, I let him loose. Now he too grabbed the osier branches, and in a few whispered words I told him of our predicament.

Our situation seemed to become more desperate every minute. I was completely stiff, numb, and so weak that it's still a mystery to me that I was able to hold onto the branches so long. I felt my remaining strength slipping away. I saw black spots in front of my eyes. I didn't know the time, but the sun was beginning to rise from behind "our" shore, shining straight into our eyes through the hanging branches. Aram, despite his dark, healthy complexion, was as pale as a corpse. Smirnov and I held him up, as the branches were slipping out of his hands. After a while, to my great relief, he seemed to recover a little.

In whispers we began discussing our meager chances of escape. I came to the conclusion that we could not count on help from our side, that if we were to be saved we'd have to save ourselves. Smirnov wisely observed that if only some firing began, it would distract attention from ourselves, particularly that of the dangerous machine gun above our heads. Our only hope lay in waiting for such a moment.

Thus we hung between life and death. For how long? I have no idea. I had lost all account of time. I knew that we could not hold out much longer. I began trying to pull myself up on the bank so as to attract attention from the opposite shore. Holding on with one hand, I waved frantically with the other. I managed to keep this up for perhaps half an hour, with no result. Then Aram took over. After some time we noticed, directly opposite us on the farther shore, a green flag, then a red one, waving violently, I suspected that these were Morse signals from our men. None of us, alas, knew anything about that code. Moving out of the bushes, I started waving again in an attempt to show that we did not understand. Somehow they must have realized what I meant, for the flags were now waving without any conventional signals. We then watched them run up a large red flag on a tall aerial wire. I couldn't figure out what was going on. I remembered that on the shooting range they used to hang out a similar flag when they were going to fire. The flag signified danger. On the "Cease Fire!" the flag was taken down. Maybe they were trying to indicate to us the moment when we should try to escape. It turned out that my reasoning was correct. Up the river, three hundred meters from the Russian side, we heard heavy firing from both machine and antitank guns. Then the same kind of barrage started down the river. Soon the Germans were retaliating. The machine gun above us gave a few long bursts, then stopped. After a while it broke out again, but much further away. The crew must have changed the gun's position. This was a tremendous advantage to us. Our hopes soared.

A roar of gunfire spread throughout the entire sector. Artillery began shooting at our shore. I carefully watched the red flag. One moment it was raised to the top of the mast, the next it was lowered out of sight. This procedure was repeated. Taking it as a sign for us to leave, I called Aram and Smirnov and together we jumped away from the shore. In a few movements we were in the middle of the river. Feeling the ground under our feet, we waded through the water as fast as our strength allowed us.

We had almost reached the beach when the Germans spotted us. First one, then another machine gun opened up. Instantly all my fatigue and numbness vanished. Already on the beach, I tore along, zigzagging. Suddenly, right in front of me, Smirnov spread his arms and fell back straight into my hands. Aram grabbed him by the feet, I by his shoulders, and between us, jumping now left, now right, we managed to get the wounded man and ourselves under cover of some bushes. We stopped there only a few seconds.

With a last effort, we stumbled on till we reached the trench occupied by our infantry post. As we let ourselves fall, literally collapsing from exhaustion, into the friendly arms of the infantrymen, one last fusillade whistled over our heads. Smirnov lay on the ground, blood gushing from his mouth. Two bullets had caught him in the back and penetrated his lungs. Aram and I could barely breathe from the pace we had been going. I felt a horrible sensation of weakness; everything went black in front of my eyes. I lay flat in the trench and raised my legs. For a moment I felt slightly better, then I lost consciousness.

I came to in a dugout. Aram was lying beside me, asleep. I found myself wrapped in blankets, without my uniform.

"Well, how d'you feel, Fred?" I heard someone ask in Polish. I looked up. Henry, the "pharmacist," was bending over me. This was the regimental dressing station. According to Henry, we had been away twelve hours, seven of which had been spent in the water. Smirnov was in hospital, alive. His wounds were not so serious as had been feared. From our patrol of ten men, only the three of us had returned. The others had either perished or been caught by the Germans. Aram and I then slept for fourteen hours.

★ ★ ★

CALM until now, the Bessarabian front suddenly became an inferno. Constant air bombardment was disorganizing the rear lines, and within two days of the German offensive the situation on the river had grown very serious. The Germans and Rumanians had already crossed it twice, each time to be forced to retreat with losses. But the third time they had broken through for good. On top of this came the sad news that the upper Prut had also been crossed. In Poland, Lwów, and Tarnopol had fallen.

Though the retreat to the Dniester defense line had gone well the first day, on the second it changed into a real panic. Utterly disorganized, all formations merged in a headlong flight. The sky was never clear of German planes, and there was little opposition. Losses from the air were colossal. Our regiment, fortunately in the first wave of the retreat, had been assigned the task of fortifying the defense line on the Dniester. The roads were jammed. I drove my truck through them, mercilessly cursing everything that got in my way.

About thirty kilometers from the river a heavy bombardment started. Abandoning the trucks on the roadside, we took to the woods. On returning, nothing was left to curse. A line of forty trucks was burning gaily. I rushed to mine just in time to rescue my gas mask, or rather what was left of it. I don't know why, but my first day at the front I had thrown out the gas mask and absorber, keeping only the mask bag, which I filled with crackers, bacon, and sugar. This strange "mask" went with me through the whole campaign, rendering me priceless services.

I continued on foot, delighted. During raids I didn't have to worry about the truck. I simply ran to where it seemed safest. Not that it was safe anywhere. It was just a question of luck.

After crossing the Dniester, I ran into some of our traffic regulation posts which were waiting to direct the columns of trucks to the regiment on the new defense line. Seeing me and a group of other drivers on foot, they gaped in surprise and told us where we could find the regiment.

I found a sad atmosphere in the battery. During the retreat twenty-five men had been killed—among them Dunin. In our gang Julek had been severely wounded, Jasiu killed. Yes, this was war, all right. The night on the

Dniester was made a hell by the roar of cannonade and explosions. On the other side of the river the enemy had cut off masses of Russians. In the morning we could already see the Germans on the opposite bank.

Next day all was quiet. Then, at noon, in the wood where we were stationed, a commotion started. At first we couldn't believe our ears. Those of us who had completed secondary education were to be transferred immediately to an officers' school at Dniepropietrovsk. In three hours the Ninth Battery ceased to exist.

Fifty-eight of us, and fifteen Russians from other batteries, all under Karpyenko's command, marched out of the forest down the dusty road, heading south. Aram, all alone, waved his hand at me for a long time. My throat was contracted with sorrow; I was losing a really faithful friend.

The following day, much to our surprise, we found ourselves loaded on freight cars, traveling toward Odessa. Moving pretty fast in spite of frequent bombardment, we reached the city at night. Poor Odessa—it was already quite destroyed. The station was in ruins. We slept the rest of the night in the park. Once beautiful and popular, the park was now packed with refugees from the west. I say popular, because this park had played a big role in the life of the Odessan garrison: it took care of the army's sexual problem. Prostitution, officially forbidden, and according to official statistics nonexistent, flowered in Odessa. In this very park the Soviet daughters of Corynth made "free" love for a meager five rubles. That's how the park came to be known as *piatiorka* (piat—five). The park also had the reputation as a place from which one could be transferred "for five to five"—"five" being the name given to venereal disease clinics of all military hospitals. Today, however, nobody "sinned" here. The population was far too busy trying to steal a minute's sleep between air raids.

Next morning we returned to the freight cars and set off on the familiar route to Dniepropietrovsk. We were kept awake by constant alarms, some necessary, others not. The country was destroyed; we saw signs of bombardment all along the line.

Approaching Kirovograd, we were caught in a heavy raid in which ten of our men were killed. At every station old peasants scrutinized our faces, in search of their boys. In front of officers and without any inhibitions, they loudly cursed the war and all those who had assured them for years that in the event of a conflict nothing bad could happen to the citizens of the USSR. They openly execrated the government, which for years had asked them to make every sacrifice in order to put the army on the highest possible level. They were indignant at having been fed on statistics which insisted that Russia's aircraft was the strongest in the world, that it had the greatest number of tanks.

"Where are our planes?" they kept asking. "Why don't they defend us?—How is it that the Germans advance and nobody stops them?"

Old women, wiping their tears with their sleeves, called out:

"Let this war alone, sons. Let it burn out. Come to us, to our huts. Put on civilian clothes. The Germans will come; they will not eat us, and we will have peace."

Such sentiments, particularly in the country, prevailed until the Germans crossed the river Dnieper. Only then did Russian patriotism and attachment to the earth make these same men and women beseech the soldiers to clench their fists and fight for every inch of Mother Russia's soil.

It took us four days to reach Dniepropietrovsk. The last day passed without raids; the country was calm, still untouched by . war. The fields of golden wheat were burning in the July sun. The difference between the front-line picture and this one seemed almost unreal.

We saw little of Dniepropietrovsk as we marched through the outskirts. We began to climb a steep street. Then the sirens sounded. An air raid, even here! But it was only an alarm. A civilian told me that up to now they had had only two short raids, which had done very little damage. At the top of the steep hill we saw a row of Red Army blocks. A gate was opened and we marched into a huge courtyard and sat down in the shade of some trees. A sergeant appeared and explained that the courses offered in this training center were for noncoms and officers, technical courses for instructors, etc. Everything was under the command of general H.Q. of the "southern front." Lolek soberly asked about the food. The sergeant smiled.

"The food is good, particularly in the schools," said he. Lolek grunted his satisfaction.

We were given quarters in one of the last blocks—large, light rooms cleaned by servants who even made our beds. Behind our block there was a wire fence. This fence surrounded the prison of NKVD. Along the fence, at every fifty meters, stood sentry towers armed with machine guns and searchlights.

"Funny," remarked Jan, "how barracks are always nearest to prisons!"

The day after our arrival Karpyenko left, after having taken leave of us affectionately and wishing us a prompt return to the regiment as officers. The commander of the school was an elderly, grayish *kombryg* [brigade commander], with a pleasant, intelligent face. He treated us to a short speech.

". . . the first of the peoples liberated' by the Red Army, you have the opportunity of becoming officers of the only democratic army in the world!"

The training began at once. During the next two months we were to receive an intensive course and graduate as lieutenants. We expected to be taught something new, but the training started off with—saluting and the parade step!

"In order to set a proper example to his inferiors," the captain informed us, "an officer must be particularly good at saluting!"

"This is sabotage?" screamed Ludwik after our first drill. "The army is badly in need of officers. It gives up hundreds of front-line soldiers to be trained fast as valuable officers, and these rascals here begin that training by teaching us to salute!"

Ludwik's attitude toward war and our part in it was admirably honest. After three days of stupid drill he couldn't stand it any longer and reported to the *kombryg* on "business that could not suffer any delay." The *kombryg* received him. Ludwik, with characteristic violence, explained what he thought of this program of training, protecting himself with the impression that the *kombryg* had probably not been informed by his inferiors as to the subjects we were "studying." The elderly gentleman listened to Ludwik, thanked him for this honest approach, and assured him that he would look into it.

During the next few days nothing changed. Ludwik was as mad as a tiger.

"Pure sabotage," he growled, "If we'd stayed at the front we might have been of some use."

One morning while we were doing rifle drill, the *kombryg* appeared. The young officer conducting the exercises reported to him. The *kombryg* called Ludwik, put his hand cordially on his shoulder, and began to walk him around the yard. We watched Ludwik gesticulating violently, waving his arms. He took off his cap and threw it on the ground. To our amazement and delight, we saw the *kombryg* bend down, lift the cap, wipe the dust off it, and calmly put it back on Ludwik's head. Ludwik, completely unaware of what had happened, was still shouting and gesticulating like a windmill. Soon after this incident the *kombryg* gave us a two-hour lecture. Ludwik, with a somber, sad face, sat on the platform while the elderly gentleman talked about the tactics of coordination of different branches of service. He spoke to the point, intelligently and interestingly. All of a sudden he broke off his speech, smiled sadly, and went off in the direction of the barracks. We stormed questions at Ludwik, who waved his hand in a sign of resignation.

"This is a decent man, even quite bright," said he. "He mourns the things going on here more than we think. However, he cannot do anything against instructions and programs that are worked out by the training division of the army staff. He agrees that they are probably thorough morons, or saboteurs, but he cannot help that. He knows that this type of training leads nowhere. He has reported it to the staff with other plans of his, but with no result"

We went to the barracks for dinner. Ludwik seemed engrossed in thought. After dinner he announced with great ceremony that he had radically changed his attitude toward war and his part in it.

"If Russians on the staff behave like this," he said bitterly, "if almost all the men we see around here are just trying to avoid being sent to the front, why in the name of hell should we be enthusiastic for a cause that is not even ours? Just because the battle is against a common enemy? Or do we want to thank them for their persecution in our country? I tell you that if they want to come out alive from this adventure, so much the more do I. I'm tired of risking my life and playing the part of a hero and exemplary citizen. From today on I'm going to be Coward Number One, and my motto: Stay alive!"

We were of the same opinion, now more than ever, because until that day Ludwik had been the only one of us to express an honest attitude toward his duties as a defender of the USSR.

★ ★ ★

IT WAS the middle of August. German planes now came over every night, the raids growing more and more severe. News from the front was confusing. The official communiques always contained the same phrases: "Our armies, having inflicted on the enemy heavy losses in men and equipment, have retreated to more advantageous positions, thus shortening the front line." Names of towns or the exact position of this "front line" were never stated. This "shortening" of the front sounded suspicious, but, cut off in the barracks, we had no idea about anything until one day we heard an artillery cannonade in the distance. At night the noise was so loud that we figured it could not be more than forty kilometers away. Our training was sped up a little, but we were still not learning anything new. We were informed of the regiments to which we were to be sent as lieutenants in a month's time. Jan and I were both assigned to the 173rd Regiment of PAP (mechanized artillery), which during peacetime had been stationed in the Crimea. We could not believe, however, that we'd go on sitting in the school for one more month.

We were right. One night the cannonade drew much closer, and the raids were continuous. We were sitting in the barracks cellars, which had been transformed into shelters. The walls shook constantly from explosions and our ack-ack, a battery of which was stationed near by. In the morning all the schools—about two thousand men—were evacuated. The *kombryg* announced that the Germans were threatening the city and that by order of the high command we were to take part in the defense of Dniepropietrovsk.

During the afternoon, trucks loaded with rifles, tommy guns, ammunition, and hand grenades drove in. We were formed into a number of groups, with a couple of officers assigned to each. Our group was composed of forty men. We left the barracks at dusk. A few kilometers outside the city we met some guides who proceeded to lead us mysteriously through the darkness. We had no idea where we were going. Suddenly the comparatively quiet night resounded with violent machine-gun fire. We didn't think we were so close to the line. The familiar rattle of a Spandau sounded too close for comfort. Fortunately, our guide, an officer, thought so too, and ordered us to disperse and take cover.

"Here we go again," muttered Ludwik, collapsing beside me. "But this time it's worse. We're infantry."

Near by, a parachute flare was shot into the sky. Immediately the night became as bright as day. I lifted my head, pressing the helmet over my eyes. In front of us we could see rolls of wire and the trenches of our defense. The territory was good for defense. On the enemy side it was flat and open; on ours hilly, affording many hiding places. The moment the flare died out, our commander began calling off our sections one after the other, assigning each to a different spot. When our turn came, Ludwik, Jan, Walter, Lolek, and myself, led by a soldier unknown to us, took over a small trench with barbed wire on the rampart. In a corner I found a heap of blankets, still warm. Our task was to keep watch ahead and fire at any suspicious move. At dawn we discovered to our right and left trenches similar to ours, manned by our boys. I didn't like the look of things. Among the five of us, I had a ten-round automatic rifle, Ludwik a tommy gun, and the others ordinary rifles. Around our waists we also carried grenades and pistols, like typical revolutionary soldiers from the illustrated papers of the day. For men on the front line whose job was to stop the impetus of an attack, such weapons were just a bad joke. It's true that our rear was guarded by some small antitank guns and a rather more regular line of firing trenches, but they must have been five hundred meters behind us. The officer inspecting our posts explained that our line had no special significance, that its purpose was simply to delay and mislead the enemy. If an all-out German attack were launched we were promptly to retreat beyond the next line.

"So I'm to delay the German advance with my own skin!" said Lolek indignantly. "Who's going to tell me whether it's an all-out attack or just a joke?"

Ludwik was beyond indignation.

"The moment they make one move," he announced, "I'm taking to my heels!"

For three days we sat in those trenches while the artillery hammered at us like mad, causing massacre among the men. Our trench was holding well. On the third day Walter had to drop out from a wound in the arm. That evening we were given bottles of gasoline, the so-called "Voroshylov cocktail," which we were supposed to hurl at the tanks. Lolek concluded that this was an attempt to make fools of us. But that night we came nearer to being made into corpses.

At dusk, heavy fire was directed at our lines. Then an infantry detachment started to attack. As usual, we hurled them back. An hour later the second lot came, with the same result. Lolek received a severe wound in the hand and had to retire. Around two in the morning came the third wave which was also forced to retreat, but after a while it attacked once more, this time with tanks. Though these tanks were not large, we quickly realized we couldn't stop them with our "cocktails." When they were within two hundred meters, we turned our backs on them. While retreating we noticed that the majority of our men had left long ago, and that we were one of the last posts. In a hurry, we scrambled through a hole in the next line not covered by wire. A voice in the darkness called out that we were not to retreat any further, but remain on the line. There was no choice. The three of us established ourselves in a corner of a firing trench and stared at the fields lit up by flares. The tanks had advanced as far as our former trench, spitting fire on all sides. Here they actually stopped for a while. We presumed that their infantry were taking over the trenches. Then they began to advance. Opposite us two tanks, hit by antitank shells, blew up, their ammunition exploding in sheets of flame. At that moment a strong wave of artillery fire opened up on us. In this hell I was still trying to see how far the tanks had advanced. But there was no point in waiting any longer. I pulled Jan and Ludwik through the wall of smoke. We started running. I can't imagine how we all managed to escape unharmed during this retreat. We ran, tripping all the time, yet holding onto one another by the hand. Ahead, another group of men were on the run. The line behind us was obviously breaking. After some fifteen hundred meters we suddenly saw the ray of a flashlight. The group ahead came to a halt. We ran to the light as moths to a lamp. We heard voices yelling: "Stop, Stop!" Ludwik asked for an officer. Someone with the face of a fifteen-year-old boy appeared and inquired what was going on up front. Ludwik was furious.

"If you want to know, why don't you go and see for yourself?" he growled.

That stopped further inquiries. Moving on a little more slowly, we had reached the outskirts of the town when a hurricane of fire overtook us. A shell hit an ammunition dump; a mass of flames rose into the sky. Crossing a weakly fortified defense line between houses, we entered the city, which was in a state of panic. Men and women with bundles on their backs were running away in masses, all in the direction of the Dnieper. We followed them. In the crowd of escaping soldiers and civilians, no one paid any attention to us. On a square we saw a military kitchen. We rushed to it and devoured whole chunks of meat that were floating about in a pot of soup.

We watched a barricade being built on the street from overturned trolley cars. Jan was overcome by an attack of honesty.

"Boys," said he, "we mustn't retreat any further! We may be needed here!"

He wouldn't listen to Ludwik, who insisted there was no hope for Dniepropietrovsk and that it would be madness to sit here and get killed. I took no part in the conversation, but with great foresight poured a third mess tin of soup down my throat, filling my gas-mask bag with biscuits. The others decided to stay on this barricade and report to the first officer they saw.

It was quite clear that the Germans had now entered the city. A major and some men were trying to fortify the barricade. On it they had placed three little antitank guns and a few machine guns. The result looked very funny.

Then heavy raids began again. The Stukas were diving over the city. We could hear our artillery firing, but from a distance, no doubt beyond the river.

Somebody told us that the river was fifteen hundred meters away, but that all three bridges were now lying in the water. The only bridge still standing was a military pontoon which was under constant fire from the artillery and the air. This was gay news, indeed.

At eight o'clock in the morning we felt a warm wind and the smell of houses burning. Ludwik climbed into the window of a house near our barricade. Suddenly he jumped down, threw his gun over his shoulder, shouted at us: "The tanks are coming!" and fled in the direction of the river.

Against tanks our barricade hadn't a chance of lasting one minute. We took to our heels. We'd barely covered five hundred meters when behind us we heard the roar of machine guns and antitank guns. Then a torrent of terrified people, still on the barricade, fell into the street.

Now began a panicky race to the one bridge. The wave of people sucked us in, dragged us with it. From side streets more masses poured in. The panic assumed the most terrible aspects. Men and women fell, got trampled on, had their arms and legs smashed.

Behind the mass the tanks were coming, firing mercilessly at the fleeing crowds. Then a frightful thing happened: with an appalling crash one of the burning buildings toppled over into the street, burying dozens of people in its wreckage. The mad wave of men, women, and children rolled up like a serpent and dashed into a side street. I concentrated all my attention on not getting crushed. Hardly touching the ground, I was carried along by the crowd. At last the street ended and we were pitched out into the open. I saw the river Dnieper.

The pontoon bridge, loaded with the struggling human mass, was burning and smoking in several places. For a moment I caught sight of Jan's face flying past in the mob.

The crowd, spilling over in all directions, tried to narrow itself down so as to get onto the bridge. Pushed from the rear, many could not manage it. Some fell onto the boards of the bridge, some into the water. The panic was increased by a fresh raid from the Stukas, whose machine guns showered the crowd with bullets. Raised by the mass of humanity, I lost ground under my feet and was carried along. I tried to figure out whether I would hit the bridge or the water. I was lucky. With a great shout we hit the bridge. The first two pontoons were on fire. I was borne through the hot strip of flames and smoke on other people's shoulders, and still kept moving. As I found out later, the bridge was broken in many places, the missing pontoons replaced by boards thrown over empty space. However, because these boards were much narrower than regular pontoons, the people who could not get a foothold fell over the sides into the water with a scream. But the loudest screams came from behind me. Turning my head and looking back I understood why. On the shore I saw a line of tanks with black crosses on their towers. They just stood there firing at the refugees as at wingless pheasants on the ground. The hundreds who had no time to jump onto the bridge spread out along the shore, seeking refuge behind folds in the ground. The screams of the trapped or wounded were not like screams I had heard before.

I could not believe my eyes when I realized I'd reached the opposite bank. To feel ground under my feet again seemed like a miracle.

I started to run, hoping to catch up with Jan, who I felt sure was ahead of me. After a few hundred meters I found both him and Ludwik gazing with horrified eyes at the crowd streaming off the bridge. They grabbed me without a word and we continued running. The farther from this massacre, the better.

We entered a wood over which Stukas were roaring, firing blindly into the trees. But I hardly noticed them. I could not rid my mind of that terrible scene on the bridge, of the awful screams of the men and women as they were pushed into the water.

All of a sudden Jan gasped that he couldn't run any further. He collapsed on the grass, exhausted. We lay down beside him, panting for breath, like fish thrown out on the shore. Presently Ludwik sat up and pulled his knees up to his chin, seemingly lost in thought.

"What's on your mind?" asked Jan in a puzzled voice.

"Friends," said Ludwik with a very serious face, "I've just come to the conclusion that what we're doing makes no sense at all. As I've already said, I don't intend to go on risking my head for a cause that is not ours. While we have been extremely loyal, others with far more reason for loyalty have simply gone into hiding. I've had enough. What d'you say to the idea of going off somewhere into the interior of Russia?"

I stared at Ludwik. So far we hadn't considered such an idea.

"Think it over," he continued. "D'you realize that two months ago our boys numbered ninety-six, and that up to yesterday we had lost forty? Today there are probably not more than twenty of us left. We should consider ourselves extremely fortunate to be alive. But one ought not to tempt Providence too long."

Jan was silent. Like Ludwik, I was also fed up with this hazardous life for a cause certainly not mine, but this hardly seemed the moment for such an important decision.

"Listen," I said, "before we decide anything, let's get away from here. We've got to try to find the rest of our boys."

The others agreed.

We continued through the woods straight ahead until we came out on a wide highway. It was packed with soldiers on foot, mostly disorganized. The officers we met kept repeating:

"Concentration for single men and groups in Podgorodne."

Podgorodne was a small village four miles away. On the outskirts we ran into a few officers with papers in their hands.

"What regiment are you from?" one of them asked Ludwik.

"From the officers' school in Dniepropietrovsk," he replied.

The officer raised his head with interest.

"Report at the large hut at the end of the street," he said, pointing. "How many were there in your school?" he asked.

"In all, two thousand men," Ludwik answered.

"Then I'm afraid there are very few of you left," said the officer sadly.

Our entrance caused quite a commotion in the hut. The first person we noticed was our *kombryg*, sitting at a table. He looked very tired. Upon seeing Ludwik, a smile illuminated his somber face. He put out his hand.

"I have thought of you repeatedly," he said warmly. "I am glad you are here."

A filthy figure with his arm in a sword belt jumped out of a corner, and with joyful howling, threw himself around my neck. It was Lolek. The night he had been shot in the arm, he had had it dressed, and then, instead of waiting for the Germans in the city, he had, as he put it, "individually forced the Dnieper, retreating to a more advantageous position, in this way shortening the front line." Seeing us whole and in good health, he was mad with joy. As Ludwik had suspected, only eighteen of our boys had survived; the rest had either perished or fallen into the hands of the Germans.

That evening news came that we were to continue to defend the Dnieper. The sector command was to organize temporary independent divisions and place them on the broken line of the river. We spent the night in a barn, undisturbed by the Germans. In the morning Ludwik woke us up. He seemed very excited.

"I've just been out and I have some news," he whispered. "If you want to run away to the rear, get ready now. A transport of five hundred men is going to Stalino to pick up some guns. The commanding officer has agreed to take us along as gun specialists. There's one condition—nobody is to know that we are from the school. Who's coming?"

We thought about it for a while. Stalino was in the Don Basin, far away in the east. Perhaps it was wise to go. Jan scratched his bearded chin.

"And what then?" he asked.

Ludwik looked at him.

"We shall run away," he said.

Lolek got up, rolled up his coat, fixed his cap.

"I'm going with you, Ludwik. Come, boys, don't let's separate!" he said in an almost begging tone.

Jan shook his head.

"I'm staying. What will happen will happen, but I will not break a soldier's oath."

Ludwik stared at him in terror.

"Man, are you mad? Are you trying to be more Catholic than the Pope himself? It's sure death."

But Jan rolled over in his coat again, thus excluding any further discussion. I suspected that his decision was caused more by his innate laziness and his desire to go on sleeping than by exaggerated loyalty. I myself was not convinced by either of them. I felt attached to Jan and I knew that I should not leave him. I also knew that further discussion with him was aimless. Although lazy, he was also terribly stubborn and, once having made a decision, nothing could make him change it. Ludwik and Lolek knew this, but they also knew that I would not leave Jan. We hugged one another affectionately. Lolek, who sometimes revealed an effeminate nature, gave way to his grief in a pond of tears. We made no more efforts to influence one another. But, having gone through so much, seen so many strange things together, we deeply regretted the separation.

Finally the two got up and left the barn. Against the grayish sky I watched their silhouettes disappearing down the road: the tall, slightly bent figure of Ludwik striding along energetically, and the stocky Lolek, rolling rather than marching on his short legs. Before I'd had time to decide which of us had taken the wiser step, I had once more fallen asleep.

★ ★ ★

NEXT day Jan and I were assigned to a group of fifty men and ordered to occupy some posts on the outskirts of Podgorodne. In front of us we had freshly dug, zigzag antitank ditches, and beyond them, behind a low forest, ran the Dnieper. Podgorodne was situated in a turn of the river, none too pleasant a position at the moment. Fire from artillery and mortars fell on this damned village not only from the front, but also from the flanks. Sometimes we had the impression that the Nebelwerfer shells were being fired from our rear.

In the provisional H.Q. of our four contingents there was a young Rumanian called Michael, from Bucharest. When the Russians occupied Bessarabia in 1940 he was in Tchemauty, and at the outbreak of the war he had been drafted. Then a third-year student of medicine, he was now a medic. Tall, handsome, with raven black hair, he went around the trenches, calmly helping the wounded. We became so friendly that he spent all his free time in our trench; he even slept in it, though he could have slept more peacefully in the H.Q. hut in the village. For some days we had sensed that he had been wanting to tell us something, but couldn't make up his mind. One evening he beckoned us into a corner of the trench. He had an understanding with a young peasant woman in the village, he said. This girl had promised to hide him, and possibly us too, in case the Russians had to withdraw. She would keep him until the Germans took over. On our sector, across the river, Rumanian regiments were fighting on the side of the Germans. Michael, whose father had an important position in the present Rumanian government, swore that the moment he got into civilian clothes and told the first Rumanian officer his name, we'd all be saved. We could get to Bucharest in no time, and once there his father would make it easy for us to do whatever we liked: go home to Lwów or Cracow, escape to Hungary or Turkey, and from there into the wide world. The proposal, though astounding, seemed quite plausible. Michael looked like a serious man, and not likely to exaggerate his father's magic influence. The problem of hiding in a cottage cellar just before the Russian retreat was not difficult to solve. This time, to my surprise, Jan consented to the idea. His desire to escape, I presumed, had at last outgrown his laziness. As for myself, I had some strange presentiment that this was a dangerous game. Though I trusted Michael, I was against the adventure for reasons that I could not, and still cannot, put into words.

The following afternoon a deadly fire of artillery and mortars was directed straight at us. After two air raids the artillery bombardment calmed down, to be replaced immediately by a storm of machine-gun fire. It was clear that an attack was coming from across the river. We received orders to be ready to retreat at any moment, but to fight a delaying action so as to give the men on the front line time to pass through us. During a pause in the firing, Michael, with flushed cheeks, tore into our trench.

"I'm now going to the cottage and will prepare everything for you. I'll be waiting."

Michael had already jumped onto our ramp when I stopped him.

"Listen," I said, "if we shouldn't turn up, it means that we were prevented or couldn't make up our minds. At any rate, rest assured of our discretion."

The Rumanian simply nodded; he knew that we would not give him away, and at this moment he was probably not too concerned about our company. In his imagination he already saw himself at home in Bucharest.

It turned out that our calculations proved correct: the order to retreat came at dusk. The remainder of the destroyed crew of the first line dashed through our posts. Up front, comparative calm reigned. The Germans (or Rumanians) were no doubt organizing a bridgehead on our bank and slowly forging ahead. We abandoned our trenches in shifts and, without panic, fell back. Jan, quite charmed by the idea of desertion, maneuvered things so that we were last in our group. Among a few cottages at the edge of the village we saw the one that was to bring us salvation. Jan looked cautiously around. All was clear. There was nothing to stop us jumping into the cottage and vanishing. I felt my heart beating violently. I couldn't understand the reason for my nervous state. Suddenly I made a decision. I pulled Jan by the elbow and dragged him into a side road, avoiding the cottage where the Rumanian was waiting for us. I didn't say a word. Jan tried to object, but he must have noticed something in my face warning him that words were useless. We ran on through the darkness until we caught up with the retreating groups. Crossing the fortified line, we entered a small wood where we were allowed to rest.

We lay down next to each other on the moss. I could sense Jan waiting for an explanation.

"Don't be angry, Jan," I said at last. "It's quite possible that I was wrong and that we have missed a golden opportunity, but something inside me wouldn't let me take this step. Occasionally my intuition is excellent."

To this Jan had nothing to say.

The night passed quietly in the wood, but in the morning shooting began again. At first we could not understand what was going on at the front. Then the wounded who were being carried to the rear told us that a counterattack by our first line had pushed the Germans and Rumanians back to the river and that if they didn't cross the Dnieper pretty quick they'd be killed to the last man.

I thought of Michael, and although he didn't say a word, I knew Jan was thinking of him too. Around noon the shooting on the river calmed down. But again German mortar fire broke out. The river bank was ours and once more we were led to our posts. We entered Podgorodne, which we had left during the night, never expecting to see it again.

Most of the houses were in flames; the village was a shambles. Suddenly I found myself trembling. From a tree at the edge of the village, on a long noose made of telephone wire, hung a corpse in civilian clothes. German work, I thought. An instant later I recognized my error: from the end of the wire, swinging slowly to and fro, hung the body of Michael. I felt someone squeeze my elbow. I turned. It was Jan.

A few soldiers were standing under the improvised gallows. Passing by, I asked a sergeant who this person was and why he had been hanged. The sergeant made a furious face.

"That son of a bitch was in our medical corps. He stayed here last night waiting for the Germans. In the morning we caught him in a cottage with two German and three Rumanian officers. They were so drunk they hadn't even heard our counterattack."

I looked at the sergeant and nodded.

"Yes," said I, "I've seen him around too!"

Jan was still squeezing my elbow. We covered a few hundred meters, throwing ourselves on the ground every few seconds because of the dense fire from grenades and mortars. Only on reaching our trench, the same trench we'd abandoned only twenty-four hours before, did Jan speak:

"Your intuition was excellent, Fred," he said. "Had we gone with that Rumanian we'd be with him still."

★ ★ ★

THE following two days we spent under furious fire. We kept hearing rumors that the Dnieper line had been broken twice—at Krementchug and at Nikopol. This news, however, had no basis in fact, since the line was holding out, even seemed to grow stronger.

The second evening an officer sent me to the group H.Q. with a written report. I had walked some five hundred meters from the trench when a violent mortar barrage started pounding. Between one salvo of shells and another I jumped up and ran, trying to reach the road where a ditch running parallel to it offered some protection. Suddenly I heard the whine of a falling mortar shell, and a tremendous wind from the explosion hurled me to the ground. By a miracle I landed in an anti-shrapnel ditch near by. I had fallen into it headfirst, and with such violence that for a moment I was knocked out. On coming to, I felt a slight pain all over my face and hot blood filling my eyes. Only later did I realize what had happened: near the ditch stood a gun carriage with a long iron rod sticking out of it. The carriage, thrown by the explosion, had tipped over and the rod fallen into the ditch. Fortunately the heavy rod, which could easily have crushed my head, had only scraped me. My lower lip felt as though it were hanging down to my chin, my nose was bleeding badly, and I realized that I was stone deaf. I did not feel any pain. With great difficulty I dragged myself out of the ditch, wiped the blood out of my eyes. But the moment I tried to stand up I fell. Terrified by the deafness, I again jumped up to my feet; an instant later I was lying flat on my face. This unpleasant process I repeated several times. Then, from my skimpy knowledge of anatomy, I remembered that the sense of balance is somewhere in the arches of the ears. Realizing that I must be quite close to the dressing station, I began crawling on all fours toward the road. Wiping the blood off my face, I let out a scream, but there wasn't a sound.

Suddenly, not more than fifteen meters ahead, a pillar of fire and smoke rose into the air and clods of earth splashed all over me. Still there wasn't a sound. Finally I saw a rag with a red cross tied to a pole. Somebody caught sight of me from a distance and ran up to me, held me up with one arm, put his other arm around my waist, and led me to the hut. While a male nurse dressed my head, I explained to the doctor that I could not hear. Only from the look in his eyes could I tell that I was speaking and that he could understand me. They placed me in a corner of the hut and I fell asleep from sheer exhaustion. After a while I was wakened, carried outdoors, and put on a horse-drawn carriage with three other wounded soldiers. Two medics got in with us, started the horses, and away we went. I began thinking of Jan, all alone in the trenches.

We rode all night, until noon of the following day. Although the carriage shook like mad on the bumpy road, I managed to sleep through most of the journey. To my great surprise we didn't have a single air raid. Around noon we reached a hospital in the middle of a forest. They laid me on a bed in a ward full of wounded men. When I asked one of the medics where I was, he produced a pencil and wrote "Novomoskovsk" on the wall. So I was not far from the polygon from which I had started out to the front two and a half months ago! With this thought, I fell asleep again.

My chin, lip, and broken nose began healing fast, but I was still deaf. I wandered about the corridors of the hospital, holding onto the walls and often collapsing on the floor. I looked for some of my comrades in the wards, but found no one I knew. When I asked the doctor about my hearing, he just nodded reassuringly. I was beginning to lose hope and kept wondering what they would do with me, now that I was an invalid.

One night, after I'd spent almost two weeks in the hospital and the stitches had been taken out of my face, I woke up. I was about to pull the blanket over my head and go on sleeping, when I thought I heard a gentle, regularly repeated sound. I sat bolt upright in bed. The ward was deep in sleep. I followed the sound with my eyes. On the wall opposite,, above the door, hung a big pendulum clock. The pendulum moved regularly, phlegmatically, and at every movement I heard a tick. I yelled out loud and trembled on hearing my voice. I could hear! A nurse rushed in from the corridor, terrified by the scream. She ran up to my bed, where I just sat, roaring with laughter. The wounded men woke up and realized what happened, since I had now jumped up and was walking among the beds, still a little awkwardly, howling all the time: "I can hear! I can hear!" They explained to the poor nurse that I hadn't gone out of my mind, but had just regained my hearing and equilibrium. I finally went back to bed. When the light was put out, I covered my head with the blanket and talked to myself, whistled my favorite tunes, rejoicing over my regained hearing.

★ ★ ★

NOW that I was well on the way to recovery, I spent my time wandering around the hospital, waiting to be reassigned to the division. One morning in the hospital office I learned some extraordinary news. Two days after my accident an order had come through from the general staff whereby all soldiers from Poland, Rumania, Latvia, Lithuania, and Estonia were to be transferred from the front lines to the so-called labor battalions. At this time I did not know the meaning of a labor battalion, so was glad to hear the news. I presumed I'd meet some of my comrades, also that I'd not be on the front line. This was the best thing that could happen.

My joy, however, was premature. Next day I was called into the office to get my discharge from the hospital. In place of my old uniform, torn and covered with blood, I was given a new one. While waiting for my papers to be filled out, I sat chatting with the doctor on duty. With the eyes of my soul I saw myself in paradise a thousand kilometers from the front, where air raids, guns, and trenches are unknown, where the sun is always shining and manna is falling from the sky. The soldier filling out my documents raised his head.

"*Tovarisch* sergeant, what is your branch of service?"

I was on the point of saying "driver" when I bit my tongue, remembering my experiences in that capacity.

"Artillery observer," said I.

The doctor signed my documents and handed them to me.

"*Bon voyage*," said he. "You'll have to get to Mohilev on your own wits. You'll find an artillery concentration there. Report at the assembly point."

I was dumfounded. Here was I, expecting a railroad trip to the east, an assignment to a labor battalion—and what did I get? Artillery again! I began feverishly explaining to the doctor that I was a Pole and therefore subject to the order assigning men to nonfront formations. Yes, said the doctor, nodding. Yes, I was right, but I was late. All the "new" Soviet citizens had been transferred from the front-line formations to labor battalions, but now it was too late. As the whereabouts of those battalions was unknown, I would have to go back to the artillery. It could not be helped. (This, I should add, is a typical example of how all problems are solved in Soviet Russia. When a regulation was issued they threw themselves at it in fury to carry it out, but only so long as it seemed new. When slightly stale it was slowly forgotten until it disappeared into oblivion, despite the fact that its execution should have been continued indefinitely.)

Cursing my sad lot, which sooner or later would surely lead me toward a miserable end at the front, I took half a loaf of bread, bacon, and sugar. I even talked them into giving me a pair of new boots. Then I walked out of the hospital gate. I wonder whether there exists in the world another army which lets a soldier out of a hospital, giving him a piece of bread, telling him to reach a town one hundred kilometers away "on his own wits," and "report" there for further front-line service!

The sun was warm; the wood smelled of resin. I walked a few kilometers and sat down under a shady spruce. The situation had to be thought over. If I'd had somebody with me at this moment, I'd have run away to the rear. All on my own, though, I could not decide on such a step. During these last two weeks I had lost a realistic vision of the situation. Today I didn't even know what the front line looked like. If only I'd had some idea as to the whereabouts of Jan, Ludwik, and Lolek, I'd have risked not reporting to the concentration at Mohilev, and tried to find them instead. Failure to report after three days equaled desertion and meant a bullet in the head. This was distinctly written out on my assignment card.

There was no choice. I had to go to Mohilev; and I even had to hurry, as three days, considering that the whole trip might have to be made on foot, was none too long for one hundred kilometers. Until the evening I took side roads, getting information from the peasants. I instinctively avoided highways which were as usual under constant bombardment from the *Luftwaffe*. On the second day I managed to get a forty-kilometer ride on a truck. I spent the night in the house of an old couple whose four sons, the day war broke out, had been in some garrison near the German frontier in Poland. I tried to convince them that their boys were all right, although I felt sure by now they'd have been killed or taken prisoner, not an enviable position. We'd already heard stories of the brutality meted out by Germans not only to prisoners of war but to the civilian population in occupied areas: the murders, tortures, and violence, the turning of all Russians into slaves on the assumption that they are an inferior race. We had heard these stories since the beginning of the war, and we never doubted their truth; but it took a long time for such excesses to turn the Russians against the Germans.

The effect of official propaganda during the previous two years of professed friendship between the two countries must not be underestimated. During that time the Russians had heard nothing about the Germans but their chivalry, valor, and friendship toward Soviet Russia. The previous years of anti-Fascist propaganda disappeared like smoke. Instead, from one day to another, the Russians heard Molotov say that "Fascism, after all, is a matter of taste." Since the outbreak of the war millions of Russians, some out of hatred of the Soviet regime, others out of hatred of war, not only thought favorably of Germans but actually awaited their arrival with eagerness, expecting help. A few months of their occupation, however, did far more than official anti-Nazi propaganda to make every Russian loathe the Germans and return to their former loyalty toward Stalin and his government.

On the third day I found myself on the banks of the river Orel, ten kilometers short of Mohilev. I discovered that I'd been lucky and that the assembly point was between the river and the city. I reported. They took my papers and told me to report the following day to the H.Q. of the 511th Heavy Artillery Regiment in Mohilev. Resigning myself once again to the lot of a front-line soldier, I filled myself with kasha in the kitchen and slept until the next morning.

Mohilev looked like a military camp. No civilians anywhere; I presumed they had already been evacuated. The town was so surrounded by antiaircraft artillery that air raids were rare.

In the office of the 511th PAP I was welcomed with open arms. Every trained man was now more precious than gold. To complete the decimated regiments they were taking freshly mobilized men with no experience, or with as little as two weeks' training with a wooden rifle. Their ages ranged from eighteen to fifty-five. The regimental commander, a tall, energetic-looking lieutenant colonel, having listened to the report of my front-line and peacetime record, assigned me as an observer to the Fourth Battery, just being organized. I was cautiously silent about my driving past. After leaving my sack and blankets in a house where I was to sleep, I went out into the street, hoping to pick up some gossip about the front. From experience I knew that all gossip is at least one-quarter true. The most prevalent rumor was that the Dnieper front had been broken. The only consoling news was that armies from the far east and Siberia had already reached the river line and were still arriving. On the northern sectors of the front, including the Moscow area, detachments had been on the line since the middle of September. Here in the southern sector, they were just arriving. Simultaneously masses of young men were being shipped to the east, somewhere beyond the Urals, to be trained for war, and, far more important, to learn how to use the military equipment which was supposed to have begun coming in from England and America.

When dusk fell I returned to my quarters. It was a small house in a garden, surrounded by a high fence of boards which blocked the view from the street Indoors I found two sergeants and two officers. The sergeants, both with front-line experience, were ruthlessly teasing the two young lieutenants, who had just graduated from an officers' school in Kuibyshev. The candle was put out and we fell asleep.

I woke to the noise of rifle fire. I jumped to my feet and just as I was, in my underwear and boots, I grabbed my *nagan* and leapt onto the porch, where I found the two sergeants. The shooting was disorderly, but violent and increasing every second. The three of us were still on the porch when suddenly the gate in the fence was flung open and two figures sprang into the garden. At first the darkness prevented us from seeing who they were. Then the moonlight fell on one of their helmets. It was unmistakably German. The sergeant next to me swerved to the rear. I noticed that he was carrying a grenade. Holding it unlocked in his hand, he let out a warning shriek, then hurled it at the Germans. Instantly three more silhouettes sprang through the gate into the garden, one of them firing his Schmeisser blindly in our direction. I dropped into a ditch which I remembered having seen by daylight. The fall was unpleasant, as the porch was high and the ditch in which I landed pretty deep. I felt a sharp pain in my knee, but at that moment a grenade exploded and fragments went whistling over my

★ 137

head. I raised myself on my elbow, with my *nagan* ready to shoot. Two Germans were lying motionless on the ground. Of the other three, one looked done for, but a second man was leaning against the fence, gun in hand, while the third, so far as I could see, was only slightly wounded. I took careful aim at the man leaning on the fence. Just as I pulled the trigger, a Schmeisser volley spattered at me, fortunately too high. I emptied my *nagan* at the shadow on the fence. Somebody on the porch was also shooting in the same direction. We heard no more from the man at the fence. The other man, who had not been shooting, now escaped through the gate. The sergeant from the porch dashed after him. He looked very funny in his underwear. I was about to get up and look at my knee when the sergeant jumped back again into the garden.

"Run!" he yelled, and promptly tore through the garden and disappeared behind the house. I did not wait long. "Run" seemed a most convincing command. Carrying the now empty *nagan* I dashed after him, forgetting the pain in my knee. Rounding the corner of the house, a burst from an automatic churned up the sand at my feet. Before me in the moonlight I could see the white pants running like mad through a field of high corn. My eyes glued to the pants, I limped on, tripping in the darkness. Reaching the fence, I vaulted over it far more gracefully than I would have under normal conditions, and continued running through another garden. White pants, on whose instinct I had bet my life, had good intuition. He kept running through the gardens, never looking around. I didn't know whether he was aware that somebody was following him, and if he did, who it was. Suddenly, beyond a fence, I saw the river. Behind me burning houses lit up the sky. I could hear the roar of guns. I stopped, wondering if there were any sense in escaping beyond the river. White pants seemed to think there was, for he had almost reached the opposite bank.

I was standing there, hesitating, when all of a sudden, a little way upriver on my side, I saw at least a hundred men tearing across the fields and plunging into the water. Close behind them I saw the splashes of machine-gun bullets. I didn't hesitate any longer. Throwing my *nagan* away, I dived into the river. The Orel was not wide. Even though my boots pulled me down terribly, I managed to swim across. On the other side I threw myself into some bushes and emptied my boots.

After a while I joined the men whom I had seen leaping into the river. To my surprise more than half of them were dressed as I was. Some of them had even abandoned their boots before taking to the water. None knew what was going on in the city; they all had the same story to tell. They had waked up and the Germans were in the street.

Meanwhile, the cannonade seemed to calm down. A major, barefoot and in his underclothes, sent two armed men back across the river to inspect the situation. We sat there, our teeth chattering with cold, until from beyond the river we heard the two men shouting that we could return. I again swam the river and began limping through the gardens toward the house. The return journey was not so easy. My knee hurt like hell; every fence presented great difficulties. I had covered about half the distance when I recognized Sergeant "White Pants," who had a slight wound in his arm. We crawled through the last fence and found ourselves in the garden of our house. Inside the fence lay the four Germans just as we had left them. On the porch we found the corpse of one of our lieutenants, in uniform. He had fallen in the doorway, perforated by at least ten bullets from an automatic. Indoors lay one more dead German, and against the wall our other sergeant, wounded, but alive. The missing lieutenant appeared only after an hour. Since he was not too keen on saying where he had been, I presumed he had spent the last few hours up in the attic.

We went out into the street. Our medics were rushing to and fro. After they had attended to our wounded sergeant, they told us what had happened. Between the rivers Samara and Orel, the Dnieper had been crossed. Promptly some German units, composed of light tanks and motorcycles, had started advancing in different directions, counting oh the bridgehead being held. The crew at the bridgehead, however, had been pushed back beyond the Dnieper, and the motorcycle units, having no other choice, in despair attacked the Russians they met. One of these groups, some three hundred strong, had come into Mohilev and inflicted many casualties among the unprepared and surprised H.Q.'s. Then the regular units from the neighborhood had arrived and liquidated the Germans down to the last man.

★ ★ ★

ONE night after a few days of hurried reorganization, we left Mohilev and headed due north. Though we were not told, none of us doubted that the Dnieper line had finally given way. We marched fast day and night, resting for only short periods. Twenty kilometers short of Poltava, we came across freshly fortified defense lines. These lines ran parallel to the Dnieper, between the rivers Psiol, Vorskla, and Orel. The first posts were already manned by infantry. We passed these trenches and occupied normal artillery posts three or four kilometers beyond. The same evening the divisions retreating in our rear began to flow in on us. From the arrivals I found out that Krementchug and Mohilev had fallen, the latter a few hours after we had left. That night, the first German artillery fire opened up on our line. My reaction to the firing after almost a month's pause was strange. The first explosions made me terribly nervous, as if I were hearing them for the first time. At the sound of a mortar howling I immediately hit the earth. I understand that such reactions are experienced by most men after they have been wounded.

For several days nothing happened. The Germans did not advance. Nor did we retreat; we just stood at our posts. We didn't dare shoot, as we had very little ammunition. I myself sat about with the infantrymen, bored stiff, and to kill time killed lice.

Then one evening after about a week of louse-hunting, the Germans launched a regular attack. For us, of course, this meant a regular retreat. The twenty kilometers to Poltava we covered in one night. In Poltava they were erecting barricades and digging antitank ditches. We had barely arrived in the city when it was subjected to a violent and most unexpected bombardment. We consequently made a hurried exit and marched straight for the steppes. From regimental H.Q. I found out that the defense staff of Poltava considered our forces so poor in numbers compared with the enemy's that they had decided not to risk such a one-sided fight. By retreating we gained in time, for the Germans, after capturing the city, had so far made no signs of advancing. So on we marched, over the Ukrainian steppe in the direction of Kharkov. On the way I discovered that our contingent was composed of no more than three to four thousand infantry, a few light tanks, motorcycles, about thirty guns of various caliber, some horse supply columns, and one or two cavalry detachments. The entire force was under the provisional command of a colonel who had last headed an army at the time of the Revolution!

For two days we continued marching north up the Vorskla, then turned sharp east. We avoided all villages which, according to the patrols, had been completely evacuated. We now had a chance to observe the results of the National Defense Committee's order to destroy anything that could be useful to the invader. The houses, if not burnt, had their roofs pulled off; all agricultural machinery was smashed; the wheat had been set on fire and the wells filled with mud. The roadsides were lined with all types of tractors, their engines blown up, often by a grenade thrown under the hood.

By the third day it was no longer a secret that our H.Q. was utterly ignorant as to the whereabouts of the enemy's forces and our own. Nobody knew who was on our left, on our right, who behind us, or if the country ahead was still in Soviet hands. We moved very slowly, sending out motorcycle patrols in every direction. That evening one of these patrols returned with the alarming news that a small village ahead was occupied by Germans. Unable to believe it, the colonel sent out one more patrol. In an hour the tragic report was confirmed. As night fell we were standing ready, waiting for the commander's decision. Suddenly, in front of us and on both our flanks, three-colored flares—white, green, and red—shot up into the sky. Only the Germans used this kind of flare. We were confused; but when the flares were fired from our rear as well, the situation became clear to everyone: we were surrounded.

It was impossible to tell the size of these surrounding forces, but it did not occur to any of us that they could be smaller than ours. The H.Q. was terrified; some officers even suggested sending out negotiators—a practice which, according to the Soviet code of law, meant a bullet in the head. Time was passing and nothing happening; only the flares continued rocketing into the sky. Since the ring seemed equally strong on all sides, our colonel decided to try an attack ahead so that, in the event of success (in which, of course, nobody believed) we could continue our retreat to Kharkov.

The attack was prepared according to the "book." It was to start with a five-minute barrage directed at the village. Then our wretched four tanks, followed by the infantry, were to attack the village.

I was up front with the infantry, to check up on the firing of my battery. The village, invisible in the dark, was situated on a flat steppe, with no stream or river between us and it. I felt that my military career was about to come to an end.

At about three o'clock in the morning our poor thirty guns started making as much noise as they could. Their fire was well aimed. The infantry began moving up so fast that it found itself on the outskirts of the village by the time the firing ceased. We heard the usual "Urraa" and a moment later the thunder of shooting.

"They're in the village," whispered the observer of the neighboring battery. Both of us had climbed a tree from which we stood a better chance of directing the fire. Reaching its climax with a tremendous roar, the firing suddenly calmed down. There fell a sinister silence. What on earth had happened?

Suddenly I saw one of our motor bikes speeding toward us. I jumped from the tree and stopped the messenger.

"The victory is ours!" shouted a pink-faced brat, as though he had just destroyed an entire army single-handed.

The other observers and I howled with joy. In my delight I took down my flask, which I had filled with *samogon* (the cheapest kind of vodka), and passed it round.

From the rear our H.Q. trucks and the rest of the men began moving forward. The old colonel, who in his military career in the Revolution could hardly have experienced such a victory as this, came riding up in a sidecar, his smile as proud as if he were in a carriage drawn by four white horses.

We then discovered what had happened. A powerful force of five hundred German motorcyclists commanded by a twenty-three-year-old captain had advanced on its own. Reaching the village and meeting no resistance, it stopped. As we approached, the Germans, with unbelievable daring, tried to deceive us with a trick—which almost worked. They sent out a few patrols with orders to fire flares. The patrols drove around us in wide circles, with results described above. Seeing our confusion, the remaining Germans calmly went to sleep. They had calculated that we would retreat and sooner or later run into their regular forces advancing from the rear. Our barrage, and particularly our infantry attack, caught them so unprepared that they made no attempt to resist. Most of them were killed, and the rest, including the young captain, taken prisoner.

The following afternoon we entered Kharkov. Half the population had been evacuated. The suburbs had already suffered great damage from air raids. Once more we watched barricades being built, antitank lines dug. The enthusiasm was tremendous. While admiring this ardor, I myself felt definitely fed up with such experiments. I could not forget that all my surviving comrades had spent the last month safely in the rear in labor battalions. I felt it was now or never for me to get away from the danger zone. The first move in this direction, however, was none of my doing. It came about quite by chance: my famous 511th PAP was to cease to exist. All its men were to be sent to the reserve center of a newly created division ten kilometers beyond Kharkov. Everyone received a document directing him to the assigned post, but the space on the document which should have shown this new division's exact location was blank. Without thinking twice, I wrote "Stalino" in the blank space, shouldered my sack, and set off to the station to try to find a train going in a safe direction. Out of honesty, but also because I knew it would be a nuisance, I left behind my automatic. I caught an evacuation train heading east—the safest direction. On the train I joined a small military detachment which possessed one great asset—a field kitchen. None of the men asked me who I was or where I was going. I just sat beside the steaming pot, devouring peas and bacon; I also took the precaution of filling my pockets with sugar. Then, at Kupiansk, the junction for Stalino, I said good-by to my companions and left the train.

On the station that evening I made friends with an old railroad man. During the course of our conversation I happened to remark that I was trying to get to Stalino. After a while I left him and went to sleep in the station. In the middle of the night I was wakened by someone pulling at my elbow. Looking up, I saw the old man standing over me, smiling.

"Now's your chance," he was saying. "A transport bound for Slovyansk has just come in. Get ready!"

Thanking him heartily, I leapt to my feet. This was the kind of courtesy I constantly found among the Russian people. Among the hundreds of men lying asleep on the station floor, this old man had taken the trouble to find me so that I might catch my train.

I caught it by a split second. It was packed with soldiers. I found myself sitting next to a very nice captain from Kiev, who began enlightening me on the situation at the front. I had no idea that for weeks the Germans had been at the gates of Moscow. By some miracle, the captain told me, Leningrad was holding out; but Orel was taken, Bielgorod and Kharkov threatened. In the south the Germans were midway between Dniepropietrovsk and Stalino. Oh, that's bad, I thought to myself. Perhaps I'd better give up the idea of offering my services to the labor battalion and continue further east.

While we were talking the captain happened to notice a faulty accent in one of my words. Looking at me carefully, he asked me whether I was Russian.

"Polish," I told him.

"In that case," he said, smiling, "I have some interesting news for you."

And he proceeded to tell me what had happened in September while I'd been stuck on the front line. The Polish government in London had made an agreement with the Soviet government, eliminating the paragraphs of the Nazi-Soviet pact which concerned Poland. [The Nazi-Soviet pact of 1939 wiped Poland from the map of Europe. According to the London agreement, Soviet Russia was willing to return Poland its former borders.] This agreement, moreover, assured freedom for all Polish prisoners of war from 1939 and for all Polish deportees and prisoners. It also recognized the organization of a Polish army which was to fight on the side of the Allies. The pact was guaranteed by British financial aid, signed by Stalin in the name of Russia, and by General Sikorski in the name of Poland. During

Sikorski's visit to Moscow the Polish flag had been raised over the Kremlin. Organization of the Polish Army had already begun at the end of September in the neighborhood of Buzuluk and Totzkoye, beyond the Urals.

I was so startled by this news that for a while I was speechless. My face, however, must have expressed great joy, for the captain kept smiling at me.

"Well," he said at last, "all you have to do now is to report to your commander, tell him you're a Pole, and he'll have you transferred immediately to your army. In the agreement there's also a provision for Poles inducted into the Red Army. They are all to be placed under Polish command as soon as possible."

Hugging the captain, I began to laugh like a madman. Soon, I thought to myself, all my troubles will be over. Once I find myself among my own people everything will change. I will know what I am fighting for, and for what cause I am exposing my head.

My one aim now was quickly to join some unit so that I could secure my transfer. The train was too slow for me. There was fire under my feet. At Slovyansk I got out. From there, by way of Artemovsk, partly by train, partly on foot, I reached Stalino.

Stalino, the largest industrial city in the mining region of Donbas, was already empty. The last evacuees had left the day I got there. Everything that could be taken out of the mines and factories had been shipped far into the east. The storage houses were either empty or burnt. In the entire city there was nothing to eat except in the few kitchens belonging to the engineers who were completing the destruction of anything that could be of value to the Germans.

I inquired about the labor battalions. They had been here, but had left. Where to? Probably in the direction of Voroshylovgrad. Asked why I was interested in them, I answered that they were my formations but that I'd been separated from them through having to go to hospital. This explanation was sufficient.

I spent the night in an abandoned house where, strangely enough, I found some potatoes. In the morning I left Stalino on my own. The city was empty, silent, as if it were dead. I began marching east.

It was the end of October; frost covered the steppe. I intended to walk until I found some military division, then report to it as a man suffering from amnesia. After all, I'd have to explain my presence somehow.

After about two hours I was overtaken by a long column of men coming out of a side road onto the main highway. They were a queer-looking crowd. Some wore military uniforms, some both military and civilian clothes, while others showed no signs of a uniform. The men looked poor and tired. Only those in uniform carried arms, *nagans* and automatics. Horse-drawn wagons carrying shovels, picks, the sick and exhausted brought up the rear of the column. The very last member of this oddly assorted army was an officer on a horse. I was leaning against a telegraph pole rolling a newspaper cigarette as he passed by. Without looking at him, I sensed that he was looking at me, and with great interest—which was not surprising. All of a sudden, in this utterly deserted region there appears a fellow in uniform, leaning against a pole, with a pistol at his belt!

I looked at him only when he had come so close that his horse's face was almost touching mine.

"You! Who are you?" he mumbled, staring at me suspiciously. He was an elderly man, with a peasant face and the rank of major. I smiled at him stupidly.

"And who are you?" I repeated.

He straightened up in his saddle.

"*Kombat* [battalion *komandir*] of the 408th Labor Battalion, Major Gusiev!" he announced proudly.

Labor battalion? Maybe this is something for me. I thought. I produced my pay book. He began to look it over. When he spoke his tone was milder.

"You're a what—a junior sergeant?" he asked, raising his eyes.

"For two months I've been a full sergeant," I lied, without knowing why.

"Promoted at the front?"

I nodded. I don't know what made me tell this little lie, which was to help me greatly. In theory, the difference between these two ranks was minute, but in practice considerable. A full sergeant was a "reliable" man who could be trusted, whereas junior sergeant was a rank generally conferred for years of service.

"What are you doing here?" asked the major cordially.

I told him calmly that after the fall of Poltava I'd had an accident from which I suffered concussion of the brain. From that day, I told him, I couldn't remember anything, and were it not for my documents I probably wouldn't even know my name. The major nodded with sympathy.

"Where are you going now?"

I shrugged. Nobody could tell me where to report, I answered, so I decided to march straight ahead. I hoped to find an assembly point in Voroshylovgrad. As a Pole, so I'd been informed, I could be transferred to the Polish Army.

The major looked at me suspiciously.

"You forget your name, but remember you're a Pole?"

"Ah," said I philosophically, "that can never be forgotten!"

The major seemed to be brooding over something. Then he looked at me again.

"I can take you with me to my battalion," said he. "I need noncoms. You will be comfortable with me, and when the time comes you can join your own army."

I thought for a minute.

"All right," said I. "It's not a bad idea."

The major told me to join his men and report to him when we stopped in the evening.

I ran ahead and caught up with the rear ranks. At close quarters the men looked in an even worse condition than I'd thought. Many of them were barefoot, bearded, their clothes in rags. They dragged themselves along. Beside me walked a man with a mustache, dressed in the uniform of the militia. He began telling me the recent history of himself and his companions. Almost all these men were from Bessarabia, mobilized the day war broke out. They had never been incorporated into the proper ranks of the Red Army. The order about the formation of labor battalions had found them in the Ukraine, running away from the Germans. Given picks and shovels, they had begun digging defense lines and antitank ditches, often right behind the front line, frequently under artillery fire and bombardment, working from twelve to sixteen hours a day, fed not much better than inmates of concentration camps, treated like prisoners, and for minor transgressions promptly court-martialed. Though told they were Soviet soldiers, and, as such, subject to military tribunal and rights, at the same time they'd been reminded that in reality they were no better than "enemies of the people" or, at best, "unreliable citizens." The battalion, assigned to no division, hardly ever received supplies from military storage, with the result that it was forced to feed itself "on its own," by confiscating cattle, flour, etc. on the way. They had no medical care whatever; the sick were driven along on horse-drawn wagons because military hospitals would not admit these men, many of whom were suffering from dysentery. The officials, moreover, aware that illness is often an excuse for desertion, never permitted the sick to stay with people who offered to take care of them. The Bessarabian peasants—unhappy, ignorant, primitive people—bore these tortures without a whisper, looking upon them as part of their inevitable destiny. Working like beasts, desperately searching for something edible at every step, they howled their sad Rumanian songs along the empty roads of Russia.

It was interesting to see how they behaved during raids, especially when the Germans machine-gunned the roads, which was their more frequent manner of attack. Instead of dispersing, the Rumanians formed a dense group on the road. The Germans, taking them for prisoners of war or some other "enemies of the nation," never fired at them. At such times the uniformed Russian escorts would hide as best they could, very often even among the Rumanians.

The man with the mustache who told me all this was from Tchernauty. For six years he had been a member of the Rumanian Communist party. Two of these years he had spent in a Rumanian concentration camp for leftist activities. After the arrival of the Russians he was freed and became a militiaman. Now, as a reward, he had been promoted to the level of a slave driver in a labor battalion, his job being to spy on and punish his own countrymen, none of whom was a "bourgeois" or a Fascist.

We were proceeding fast, halting every two hours for a few minutes rest. At one of these stops I walked into an empty hut and attached an additional triangle to my collar so as to conform with my lie. The major hadn't had a chance to see my insignia before, as I'd been wearing an overcoat, to which insignia were seldom attached. I studied my reflection in the window: two triangles looked much better than one.

Having covered thirty kilometers, we stopped in a deserted farm for the night. I reported to the major, according to his order. He told me to sit down and offered me some vodka. Two Rumanian cooks covered a table with platters of roast chicken, goose, white rolls, vegetables, and fruit. They walked on their toes so as not to disturb their powerful master. During our conversation two more Rumanians came in. While the one took off the major's muddy boots and washed his legs in a bowl of warm water, the other prepared himself to shave the monarch, who accepted these services as if they were his due.

Though indignant at what I saw, I tried not to show it. The major was looking at me with interest, probably expecting me to make some remark. Since I didn't, he made his own:

"You can assure yourself of these same comforts in your platoon. They like it!"

"They," of course, referred to the Rumanians.

"But I don't like it," I blurted out.

Raising his surprised pale-blue eyes, he promptly changed his tone. In a harsh voice he ordered me to report to the Third Company and take charge of the platoon. I stood up, saluted, and left. I had made a mistake. I should not have antagonized him. Now he was my enemy.

I went to the Third Company and found its commander, an old lieutenant, in front of his house. He, too, invited me in, but expecting the kind of hospitality I had just witnessed at the major's, I declined, explaining that I was tired. The company's *politruk*, who had up to now been temporary platoon commander, then led me to my platoon. I took over his duties and asked him what the men were doing. He shrugged.

"Probably eating," he said in disgust, and added, "That's all they ever do!"

I went out to look for the company kitchen. I found it in one of the courtyards—three large steaming pots hung on chains over a fire, with a cook standing by. At least a hundred men were staring at the pots in silence.

As I approached, the cook asked me if he might serve the soup. I nodded. The men formed a long line, the "elder" of each section at its head so as to prevent anyone taking a second helping. With mess tins, pans, and cans, they walked a few steps away from the pots and quickly poured the soup down their throats. I say "poured," as it contained nothing to chew. Then they walked back again for the dirty hot water known as "tea." These men were clearly starved. The *kombat's* guard and cooks, on the other hand, were served thick soup with pieces of meat and potatoes. This was their privilege.

Horrified by this spectacle, I returned to my hut, which I was to share with the *politruk*. We sat over a table spread with large chunks of fried meat, bread, and milk. I ate in silence, refusing to answer the questions of the chattering *politruk*. Though still far from revolting against the human suffering I had witnessed, I was deeply shocked. I finished eating, prepared a bed for myself, and went out for a breath of fresh air. In the fields beyond the road I heard voices. Approaching closer, I found the Rumanians preparing themselves for sleep on the open stubble field. All round them were lines of empty huts, which their owners had evacuated three days before. These huts contained sufficient beds for the entire battalion, not to mention what they offered in warmth and shelter. The Rumanians, surrounded by guards, were sleeping in the cold, which at dusk turned to bitter frost. I returned to the hut.

The *politruk* was lying on the bed in his underwear, smoking a cigarette. He offered me one from his pack. I refused, wrapped myself up in my coat, and fell asleep.

★ ★ ★

AGAIN we marched a whole day. I couldn't take any interest in my duties as platoon commander. I walked like an automaton, trying to think of some way of escaping from this situation. I realized that I had made a mistake. Owing to the shortage of noncoms, they were not likely to let me go easily. I knew, too, that if I did not comply with the duties imposed on me I'd be held responsible.

When we stopped in the evening after a thirty-kilometer march, I went to the battalion H.Q. I found the major again at the dinner table, no less luxurious than the day before. He did not ask me to sit down, nor did he offer me any vodka. With his mouth full of macaroni and cheese, he tried to ask me what I wanted. As calmly as I could, I suggested that perhaps I should make an application for my transfer to the Polish Army, and asked if he could tell me to whom I should apply. The major turned on me his malicious watery eyes.

"What, don't you like it here in the battalion?" he asked.

Again I made a mistake.

"No," said I.

"Perhaps you feel sorry for the Rumanians?" the major suggested.

I shrugged.

"Well, well, mind you behave, or you'll find yourself digging fortifications too! After all, you're also a foreigner. Being a sergeant won't help."

The thought of it made me feel cold.

The major went on staring at me.

"Ask the commissar what you should do," he added in a slightly milder tone.

I felt I had to say something, if only for the sake of my prestige. "Don't forget, *tovarisch* major," I stuttered, "that the order transferring Polish citizens to the Polish Army has been confirmed and signed by *tovarisch* Stalin. I don't suppose you want to criticize His orders?"

The major jumped up from his chair. His face was gray. He tried to say something, but couldn't get it out. I stared at him coldly. Then he recovered and tapped me on the shoulder.

"What's this? Don't you understand a joke?" he asked coarsely. I was again quite sure of myself.

"I don't enjoy that type of joke," said I, taking advantage of his terror. "And to me, what *tovarisch* Stalin says is sacred!"

From a table behind me a young man came forward.

"Well, let's go into the next room," he said, turning to me. "We'll write out your application."

The young man turned out to be the battalion commissar. We sat down at a table. He offered me some tobacco and began to inquire about my past. Under his dictation, I wrote out an application to the staff of the army for a transfer. He promised me I'd receive an answer in less than ten days, since such applications were discussed at the nearest *voyenkomat*.

"Of course, you must go to your army," he assured me. "That's where you belong, and we are fighting the same enemy."

I was about to ask him what he thought of the treatment of the Rumanians, but bit my tongue. Why antagonize two men in one day?

Next day, with the enemy's front line moving up in our rear, we continued our retreat. The areas we crossed were deserted and "prepared" for surrender to the Germans. Late in the evening we reached Voroshylovgrad. The city was packed with soldiers. Some special divine care must have saved it from bombardment, for the houses were so overflowing with men that one bomb would have caused tremendous losses. Most of its inhabitants had been evacuated beyond the Don; some, however, were still in the city. Finding no room there, the battalion had to spend the night in the suburbs. Only the battalion's H.Q. remained in town. Appointing Kola, a very nice Russian university student from Kiev, to replace me in the platoon, I went out and mingled with the crowd. My mind was still intent on that transfer.

Having made sure that the battalion was to remain here for two days, I found a place to sleep and moved on in search of the *voyenkomat*. I got there late at night. When the soldier on duty asked me my business, I declared it was something which could not be delayed, that I had to see the commander immediately. The soldier disappeared. When he returned, he informed me that the *kombryg* was at the moment attending an important conference with many high-ranking officers, but that since my mission was so urgent and I had come at so late an hour, the commander was willing to make an exception. Opening a door, the soldier let me pass in front of him. I knocked at another door and, without waiting for an answer, pushed it open. I was immediately blinded by a brilliant light. I stood there, blinking. As my eyes slowly grew accustomed to the glare, I began to inspect the room. At a round table sat fifteen officers, none with a rank lower than major. The room was gray with smoke. Bottles of vodka, glasses, and snacks covered the table. Across the knees of two officers sprawled a couple of drunken women. On a wide couch lay two more women, snoring. Though somewhat taken aback by all this, I managed to report that I had come to see the commander of the *voyenkomat*. An elderly man with a fat, red face, looked me up and down.

"What is there of interest that you could want to tell me, sergeant?"

Now, my experience had already told me that not one of these men came from the front-line division. No longer young in ill-fitting uniforms, they were certainly all reserve officers—some of them probably commanders of labor battalions.

Staring at the half-drunken *kombryg*, I began to explain that the object of my visit was to seek a petition for my transfer. The *kombryg* listened and nodded, while his face began to assume a more conscious expression. My story finished, I stood waiting for an answer. He thought for a while and then asked:

"Does your commander know you're here?"

In a moment of honesty, I replied that he did not.

"Oh, yes, he does!" came a voice from a dark corner of the room.

I glanced in that direction and grew stiff; into the circle of bright light marched Major Gusiev. He was staring at me with his ugly, watery little eyes.

"So we meet again!" said he. "This will not pass unpunished. You know that without my knowledge and permission, you have no right to address yourself to a higher authority?"

I had very little to lose. Without thinking, I calmly said:

"Consider that I still haven't said a word about life in the 408th Battalion!"

Gusiev paled. I watched the *kombryg* observing us attentively. Quite conscious now, he was trying to read between the lines of our conversation. One of the women filled a glass of vodka, rose from the table, and came up to me.

"Drink, little one!" she said in a hoarse, drunken voice, pushing the glass under my nose. With a quick gesture I knocked the glass from her hand. It smashed in splinters on the stone floor. The woman let out an oath through clenched teeth.

"Oh, you son of a bitch!" she howled and, leaping at me, tried to slap my face. I grabbed her arm, twisted it, and pushed her away. She tottered, and collapsed on the couch, crying to high heaven. During this scene nobody moved or said a word. The *kombryg* looked at me with approval. He turned to the girl.

"Shut up, you bitch!" he growled.

Her crying stopped and she hid her head in her arms. The *kombryg* turned to me.

"Now just suppose, sergeant," he said, "that this woman had been my wife?"

"Your wife would not behave like that, *tovarisch kombryg*," said I. "But if she did, I'd do just the same!"

"Well, well," said the surprised officer, "you are a courageous man!"

So far, Gusiev hadn't uttered another word; he just stared at me with his hog-like eyes. The *kombryg* straightened his shoulders.

"Don't worry, sergeant," said he. "The day will come when you will be transferred to your army. But for the time being Major Gusiev surely needs you. The front is too close for us to disorganize companies."

I didn't know what to say, so I said nothing. The *kombryg* looked at me again.

"Well, go and sleep, sergeant," he said. "I'll tell the major what he's to do to transfer you as soon as possible."

I saluted and left. In the corridor the soldier looked at me inquiringly.

"Well, are they still drinking?" he asked.

I nodded.

"This has been going on every night for two weeks," he told me. Outside, passing the shuttered windows of the *voyenkomat*, I heard the sounds of a guitar and a drunken female voice howling:

Vdol po ulitzy myetyelitza mietiot—"A blizzard is howling down the street ..."

Before leaving Voroshylovgrad I searched the whole city for a sign of my comrades. In vain. All I learned was that the reorganization of the labor battalions was supposed to take place in Millerovo, southeast of the city. We made this journey partly on foot, partly by train. Our food situation was now appalling. Before taking to the road I filled myself up with the supply of salted bacon I had in my bag. It was tough and it stank, but Kola and I praised it because it was filling. The Rumanians, however, had nothing to fill them. Their condition grew worse every day. Their faces had fallen in, their eyes burned with hunger. Whenever they saw a few frozen potatoes in a field beside the road, they dived for them like animals, devouring earth, skin, and all.

To cap our troubles, the cold was now intense. A promise of winter uniforms had, like all promises, come to nothing. For two cakes of soap (saved from the hospital) and a spare razor, I persuaded a Millerovo housewife to part with three old but warm sets of underwear and a short sheepskin jacket. Wearing four sets of underclothes, two summer uniforms, the jacket and overcoat all at once, I managed to keep warm, but this wardrobe proved a bit of a handicap when satisfying certain physiological needs.

In Millerovo, of course, there was no reorganization of labor battalions. Every time I visited the commissar to inquire about my application I was given a new one to fill out; the commissar smiled politely and promised that a favorable answer would soon arrive. Though I didn't believe a word he said, there was nothing I could do but go on waiting.

From Millerovo we continued marching south, but with an assigned task—to build fortifications on the Donetz. I myself did not have to work, but I had to see that my platoon did. We got up at five in the morning and, after an onion soup, marched to the river. Every platoon was given a special job by the company technician. From six in the morning to two in the afternoon, the Rumanians—starving, frozen, their feet in rags instead of boots—dug trenches and antitank barricades. If they had finished their particular job by two, they were given more soup and continued working till dusk. If not, the technician kept them at it till four, by which time their soup was cold. On the second day I promised the technician that if he overworked my platoon I'd shoot him. To convince him that I meant what I said, every time he glanced my way I gripped the butt of my *nagan*, snarled, and made the most terrifying faces. I was always expecting him to report me to the *politruk*. But he never did.

After a week we moved downriver a few kilometers to a village called Niznyi Ruskoi Bishkin, where Kola and I found quarters in the house of a Cossack family. The old father clearly didn't like the look of my uniform. Kola, who was in civilian clothes, he treated much more kindly. When, during supper, I mentioned that I was Polish, he instantly changed his manner, and for the first time I realized how the Cossacks hate the Soviet regime. As a freedom-loving people they had resisted Soviet authority as long as they could; but soon, tortured economically, faced by the ghost of starvation, they had had to make peace. Even so, they still managed to retain some of their former privileges and customs. Cossacks drafted into the army, for example, were not inducted into any branch of the service. There were special regiments of Cossack cavalry in which all the youth from the Don region served. A farm village sending a boy to service, however, had to supply him with a horse, harness, uniform, and sword. This uniform consisted of black trousers, a shiny black tunic buttoned at the neck, and over this a cape, usually black, lined with material of various colors. On their heads they wore round black caps made of fur, and across their chests ammunition belts filled with highly polished bullets. Mounted on their horses in these uniforms the Cossacks looked superb. The village, in order to give its enlisted men all these accessories, often had to incur debts, but never gave up. Physically they presented an extraordinary type of beauty: tall, graceful, dark, with aquiline noses. There was something aristocratic in their manners which I have not found elsewhere.

In Bishkin we worked a few days, then moved on as far as the Don, stopping in the village of Kurmanskaya. Again I asked the commissar about my application. This time he was not at all polite and suggested I go to hell. I did not go to hell. Instead, asking Kola to replace me, I took advantage of the battalion's only truck, which happened to be leaving for Kotelnikovo, about 150 kilometers away. The commander of the *voyenkomat* in Kotelnikovo, while admitting that one day I would undoubtedly be transferred to the Polish Army, told me there was no use my thinking about it at present as there was a serious shortage of men.

This unsuccessful trip to the *voyenkomat* took me three days. Kola tried to cover up my absence by explaining that I was ill, but the major soon got wind of it. No doubt remembering his old grudges, he considered it sufficient reason to deprive me of my position of platoon commander and assign me to the platoon as a simple laborer. I was to go to the sector and dig ditches. I had not the slightest intention, of course, of obeying this order. As a matter of fact, I wasn't feeling at all well. Starvation had given me a serious case of avitaminosis. All my teeth were wiggling like the keys of an old piano. My legs had begun to swell and they were covered with sores. So I just stayed home. When possible, Kola brought me my soup. When he was at work my landlady went to get it for me. The people in whose house we lodged were very kind; they helped us as much as they could, but since they themselves were starving, that help was more of the moral kind, which, as everybody knows, does not relieve hunger.

One evening Kola came back from work with a frostbitten nose and tragic news. The major had been asking for me. When informed that I was still not working, he had ordered my company commander to tell me that if within three days I had not reported for work, my daily ration of bread and soup was to be cut off. Kola was to bring my decision to the lieutenant. That decision was brief: having given Kola a detailed description of where the major could kiss me, I told him to inform the major that it would be very unwise for him to let a sergeant with a front-line career work with a shovel in a battalion such as his. Kola walked out with a worried face. A few minutes later I heard someone knock at the front door. When the landlady opened it I heard a man inquiring about the "Polish sergeant." The man's voice sounded strangely familiar. Dragging myself off the bed, I went out to the porch. Who should be standing in the door but—Lolek! As fat as ever, in a worn-out coat, dear old Lolek, whom I had left in Podgorodne! Throwing his arms round my neck, Lolek cried like a salesgirl at a sentimental movie. He blew his nose loudly with two fingers, pushed my head back to look at me, and then cried again.

His situation was not too bad. He and Ludwik had arrived in Stalino from Podgorodne at the same time as the appearance of the order about the organization of labor battalions for "new Soviet citizens." They had been together for a while. Then Ludwik disappeared. Lolek thought he probably had a job in the supply service. Lolek himself had had a very hard time digging ditches, until he conceived the wonderful idea of reporting as a mill specialist. The supply service, which doled out flour to a few battalions, owned mills, partly damaged or dismantled in the process of evacuation. Lolek, as a "specialist," was to put these mills into working order. With his typical insolence, he had got one of them running and was now a member of its Diesel engine crew. The mill was located at the other end of Kurmanskaya. He suddenly produced a small bag of flour and handed it to me.

"It isn't much," he stuttered, still sobbing, "but I wasn't sure I'd find you. Tomorrow I will bring you more."

I told him my own history, and the latest news.

"Don't worry," he reassured, me optimistically. "As long as we are close you will have as much bread as you want."

I felt better right away. The possibility that some of my comrades might be alive raised my morale considerably.

Next morning Lolek brought me a large bag of flour. My landlady made noodle soup, noodles without soup, then pancakes fried on the stove. Everything was delicious, and we filled ourselves completely. In two days the bag was empty.

The following evening Kola returned from the kitchen with only one soup. The cooks had been forbidden to give me a ration. Kola was desperate; in this he saw the end of my existence. Placing the soup on the table, he laid two spoons on the mess tin.

"Eat, Fred," said he.

I was very touched by his gesture. Those who have not known hunger will probably not be able to appreciate this sacrifice. A single portion of this soup was meant to keep one man alive. Despite Kola's insistence, I could not accept it. Kola was trying to bring himself to eat the soup alone, when there was a knock at the door. A huge Rumanian called Rantuz, from my platoon, walked into the kitchen. Very embarrassed, he began stuttering in broken Russian. At last, from behind his back, he produced a mess tin covered with a white rag. Then he put a hand in his pocket, drew out a small package, put it on the table, and made for the door. I stopped him. In the mess tin I found an abnormally large portion of soup, and in the package a ration of bread. The Rumanians, knowing that I was deprived of soup and bread, had cut down their own miserable rations for my benefit, while the cooks were secretly sending me some soup. I could not refuse such a gift.

Even though I was lying in bed all day, my disease was getting no better. On the contrary, it progressed. One day Lolek arrived with another sack of flour and the news that his mill was being transferred to the neighborhood of Stalingrad, and he with it. We had known that this moment was bound to come, and we did not make a tragedy of it.

"I hope we will meet again," whispered Lolek, sobbing.

"At worst, in the other world," I answered.

★ ★ ★

LOLEK'S departure meant the end of my flour. The soup, gift of the cooks, also came to an end as the politruk inspecting the kitchen discovered what was happening and promptly stopped it. I was left with a small supply of flour and the bread brought to me daily by Rantuz. My landlady from time to time shared a piece of meat or cheese with me, but this diet could only stave off starvation or a complete collapse. By now one-third of the battalion could barely walk. In my company twelve men had died of sheer fatigue and undernourishment; there were several cases of typhus, and then others were on their backs, too weak to move. The cemetery behind the village was filling rapidly.

The day we were to leave Kurmanskaya, I was told that I'd have to go on foot or on a cart with men suffering from typhus. This was too much for me. I decided to quit. From early morning a caravan of Kalmuks [a semi-nomadic Mongolian tribe, whose republic stretches from the eastern bank of the Volga down to the Caspian Sea] had been tramping the road in the direction of Kotelnikovo. Hailing from the region of Astrakhan, these Kalmuks had been mobilized for fortification labor on the Don. Now they were going home. Loaded on wagons and sleds drawn by horses, mules, even camels, the procession looked quite unreal.

Bidding farewell to my landlady and Kola, I escaped by the gardens onto the road. To avoid being seen I jumped on the first horse-sled I encountered (I don't trust camels), and so started on a new voyage into the unknown. The people on the sled made room for me without asking any questions. Lying on my back, covered by smelly fur, I could say that I was traveling in comfort.

At dusk we stopped in the middle of a steppe. Each family put up a small leather tent and started a fire. An ancient Kalmuk approached me and, in almost incomprehensible Russian, informed me they were heading for Kotelnikovo where they expected to be paid for their work on the Don, Then they were to continue by train in the direction of Astrakhan. I told him briefly that I was sick and that I would like to accompany them as far as Kotelnikovo. Nodding, he was about to go when he turned and asked what was wrong with me. I pulled off one boot and unwrapped the rags of a sheet with which my legs had been dressed. He took one look and nodded.

"Hunger," he said. "Wait, I'll give you some medicine."

A few minutes later he returned with a leather bag. Out of it he produced some fresh fish fat with which he began dressing my wounds. Then he handed me a large chunk of smoked bacon.

"This is the best remedy," he said, his old face wrinkling in a smile.

Yes, food was the best remedy. The family—composed of the father, mother, two grown-up sons, and a baby wrapped in fur—prepared a greasy, tasty soup with meat and a corn paste. I ate so much that my stomach grew to a tremendous size. Then I lay down beside the fire and with the cold wind whistling over the steppe, I fell asleep.

★ ★ ★

I TRAVELED with the Kalmuks for three days. In spite of the frost and wind, I felt far better. The food had a magic influence on my avitaminosis.

As for the wounds under the dressing of fish fat, I could almost see them healing. Even my teeth stopped wiggling. When we reached Kotelnikovo I bade the family an affectionate farewell and dragged myself to the voyenJcomat. I wondered if I should run into the commander who had thrown me out a few weeks ago. If I did, I was done for. I walked into the hall. On the commander's door there was a card ordering one to report to the soldier on duty. Suddenly the door opened and a man came out. Through the doorway I could see a Russian I didn't know sitting at a desk. With a sigh of relief I knocked, and without waiting for an answer, walked into the room. Then I stood stock-still. At a second desk sat my old enemy, the commander who had thrown me out. I turned, stuttered some words of apology, and escaped to the hall. Behind me I heard a furious voice, calling the soldier on duty. Seeing a sign saying "MEN" I pushed open the door. In the toilet I began to think fast. I was determined not to leave the building without having spoken to the commander. Then I got an idea. I took off my overcoat, strapped my belt over the sheepskin jacket, and carefully hid the insignia under its collar. I tied the remains of an old handkerchief around my face, as if I had the toothache, pulled the fur cap down over my eyes, and looked at my reflection in the window. My own grandmother would not have recognized me. Then, according to regulations, I reported through the soldier on duty as an enlisted man with the name of Vasilev. In half an hour I heard my new name called out. I walked into the room and saluted. The commander and the man I presumed was his adjutant stared at me, speechless. Finally the commander rose from his desk, walked around me, asked me what I wanted. Without faltering, I said I was a soldier of the 408th Battalion, now on its way from Kurmanskaya to Kotelnikovo. As a driver, I had been sent ahead to fetch some bread. A Polish friend of mine in the battalion had asked me to get him some information about his transfer to the Polish Army.

Both men still stared at me with incredulous eyes.

"And why didn't this Pole come here himself?" asked the adjutant.

"It's not quite certain that the battalion will pass through Kotelnikovo," I invented there and then.

The commander rubbed his forehead.

"Some time ago we had a Pole here who asked precisely the same question," he said. "He was also from the 408th Battalion, but he was a sergeant. A cheeky fellow, a deserter, I think."

I swallowed. He was not exactly paying me a compliment.

"Must have been someone else," said I. "My friend is a private."

The commander was obviously trying to remember our previous conversation. Suddenly he banged the table with his fist.

"Yes, he was surely a deserter," he howled. "We should never have let him go!"

He now looked at me with more sober eyes.

"Tell your Polish friend to come here," he said in a milder tone. "I will look over his papers and see what can be done for him."

There was no point in my staying here any longer. I picked up my bag and coat, saluted, and walked out.

The situation was bad. The nearest *voyenkomats* were either in Krasnoarmiesyk, in the northeastern direction, or in Proletarskaya Stanitza, down the main railroad track.

I had great difficulty finding a place to sleep. Every building was full of army men. I at last found room in a house occupied by several officers and two girls in uniform. While sitting on the floor, shaving, I noticed in my mirror one of the girls staring at me. After a while she came closer.

"You from Lwów?" she asked.

I turned my head. How could she know I was from Lwów? Quite sweetly she pointed to my tin candy box from Wedel [a large Polish candy firm], in which I kept my shaving things.

"I spent eighteen months in Lwów, almost until the outbreak of war," she informed me.

We began to talk. She was an assistant at the University of Kiev, and had worked in the Lwów university library. We exchanged reminiscences about Lwów, reminding each other of its streets and shops. We discussed European literature, with which she was well acquainted. Her name was Lidia, and we got along famously. Meanwhile the other girl had been preparing a meal for the officers. She called Lidia to the table and turned to me.

"Please have supper with us," she proposed.

I did not have to be coaxed. We ate meat out of cans, something I had not seen for years, good tea, and sugar. The officers were all from an infantry regiment which had just spent three months at the front. They were now on their way to Stalingrad for a furlough and a military course.

We talked late into the evening. After the officers and the other girl had gone to sleep, Lidia and I still went on talking. Certain by now that I could trust her, I began telling her about myself and my present situation. She listened with intense interest and gave me her opinion.

"If you want to get into your army legally," she said, "try to reach a front-line unit. Don't go any further to the rear. You're far more likely to fall into the hands of the NKVD. They smell a deserter in everyone they see."

I considered this very good advice. Next morning, after studying a railroad map at the station in Kotelnikovo, I decided to go either to Proletarskaya Stanitza or to Salsk. I finally chose the latter because I discovered that a big artillery concentration was taking place in Salsk and new regiments were being organized with large supplies of equipment and few men.

Arriving in Salsk on November 25, I at last found out something about the situation at the front. On our southern sector, the Germans, though impeded by a more decisive Russian resistance and a winter for which they had not been prepared, had just taken Rostov. They were now lined up in positions parallel to and about eighty kilometers west of the Donetz. The German air bombardments had decreased to a minimum. Their gasoline supplies, arriving from the west, were being regularly exploded by partisan troops in the German rear. While German tanks were having great difficulty moving in the deep snow, their crews were freezing and falling ill from lack of warm clothing. The Red Army, on the other hand, had become more vigorous and shaken off the initial shock of the flashing Nazi drive inside Russia. Allied help was now reaching Russia through Murmansk and Archangel. Finally, the Russian armies were prepared for winter, while the Germans were not.

With this optimistic news, I arrived at the assembly post in Salsk with a lighter heart. Men there were flowing in from all over the place: from destroyed units, hospitals, and post-hospital furloughs. Standing in line, I laughed to myself at the paradox that had brought me here. On the safe "rear" I had seen the vision of death from starvation, for which reason I was returning to the front! Better to die at the front from a German bullet and with a full stomach, I thought, than at the rear from hunger!

The officer at the table was looking over my Red Army pay book. He raised his smiling eyes.

"Where have you just come from?" he asked.

I was pretty sure that he had already summed up my situation. The last assignment date in the pay book was more than two months old.

I answered his smile with a smile.

"From there!" said I, pointing over my shoulder.

He smiled even more broadly.

"Did you get any rest at the rear?" he asked, apparently realizing what had happened to me since I'd left the 511th PAP in Kharkov.

"I almost died of hunger," I whispered.

He laughed out loud. No doubt he had run into such cases before, or maybe he himself had had a similar experience.

"Well, we'll fill you up," he said. "We have plenty of food."

He handed me my new assignment card: "Artillery H.Q. of the 117th Infantry Division."

"H.Q." sounded good. It smelled of a better kitchen, and not too near the front.

★ ★ ★

THE two days in Salsk were paradise after the black days in the 408th. I received a warm, brand-new uniform, a sheepskin coat down to my knees, with double sleeves and mittens. The food was tasty and plentiful.

There were rumors of an imminent Russian counteroffensive. On the night of November 26 my division set off by rail in the direction of the front line. In the morning we received the first news: the offensive along the entire southern front had begun on November 24. With three armies, Timoshenko was pushing the Germans back. One army was heading for Kharkov, the second descending from the Donetz toward Rostov, and the third, to which my division belonged, was to attack Rostov straight across the Don. Odessa hadn't fallen till the middle of October, when the right wing of Mannstein's army attacked the Crimea, cutting off the city. By some miracle, the Odessan garrison had escaped in the battle and, having destroyed the bridges and factories of Odessa, got out to the Caucasus by sea. It was these armies, reorganized and well rested, which were driving toward Rostov.

We arrived at the Don on the night of the twenty-seventh. From behind the river thundered German artillery fire, reminding me of old times. But when you know you're on the attacking instead of the defending side, the sensation of gunfire is very different. The artillery regiments belonging to our division occupied firing posts and were shooting at German positions beyond the Don. A young officer, two sergeants, and I were on an observation post beside the river. This was a rather peaceful position, as the German fire was falling to our rear. Near us masses of infantry, tanks, and all kinds of equipment were moving forward. At night the Don was to be forced, and, with Timoshenko's two other armies, we were to enter Rostov. Around midnight our artillery fire developed into a hurricane. The thunder of the guns merged into one tremendous roar.

"Our men are probably forcing the river," one of the officers yelled into my ear.

We were lying flat, our chins pressed to the ground, listening to the roar of fire above us. Around four o'clock in the morning the radio in our shallow trench began feverishly urging us to increase our range of fire. This meant that our infantry had got hold of the bridgehead on the other side and was pushing forward.

At dawn the situation was clear. The Germans were defending themselves in the city, but without great conviction. The Don had been spanned by a pontoon bridge on which aid for our armies was going across and transports of wounded returning. The struggle lasted a whole day with a never weakening intensity. Timoshenko's second army, coming from the Donetz, was also approaching the city. At night we received the order to cease fire and cross the river. Next day our artillery regiments were to arrive and we were to direct them to the new firing posts.

The four of us got on a truck and without mishap crossed the Don and spent the morning in a destroyed cottage.

Rostov was taken. The infantry patrols were already in the city. We stood near the bridge, waiting for our units. Over the bridge came a procession of men and equipment. For the first time in my front-line career I saw armies pushing forward, organized and in perfect order. We waited for our regiments until late in the afternoon. The posts indicated in last night's order were already out-. dated. The situation was much better than even the most optimistic orders could have anticipated. In the evening, divisional H.Q. organized a reconnaissance unit consisting of forty men, under Captain Gonkov. This unit, composed of men from the infantry, artillery, tanks, and engineers, was equipped with two trucks and two motor bikes and sidecars captured from the Germans. This and several other such units were to advance as the vanguard of the army. I took over one of the motor bikes—a superb, almost new Zundap. As the machine-gunner in the sidecar I had a gay sergeant by the name of Serov.

From Rostov we took the route leading through the steppe along the railroad line to Taganrog. Twenty kilometers out of the city we caught up with the main stem of the army attacking the Germans. Passing through it and driving with all possible precautions, we reached the leading reconnaissance units and with them moved on, close to the heels of the retreating Germans. This was not a particularly pleasant or safe task. As reconnaissance units we were not supposed to show fight in the event of enemy resistance, but to wait for the arrival of forces following us. On the other hand, there was always the chance that the Germans might set an ambush and do away with us before help could come.

The Germans, fearing encirclement, continued retreating. Only on the second day did they begin to resist. By this time we were again in the role of artillery observers, and the tanks and infantry were passing through us. More and more prisoners were coming in, all in a terrible physical shape. With frozen ears and noses, wrapped in rags, they looked awful. These were the remnants of the proud conquerors of Europe of 1939, even of a few months ago here in Russia. Had they not been Germans they would probably have inspired in me a feeling of pity, but now I felt that justice was being done.

About thirty kilometers before Taganrog I witnessed a scene that I shall not easily forget. A battle was raging in front of us. To delay our advance, the Germans had left behind them a number of small groups which, doomed to perish anyway, fought to the end. Serov and I were lying on an observation post on a hill. Beneath us we could see a small village in flames. The battle had just passed beyond it. Over the radio the artillery was ordered to cease fire and push forward. Loading the radio and scissor telescope onto the motor bike, we drove slowly forward until we reached the village. Near us, on a hill, we saw four cottages about to collapse in flames. Soldiers, diving into the burning buildings, were carrying out women, old men, and children—many already dead, others terribly scorched. Serov leapt from the sidecar to see what he could do. He came back with a furious face, his teeth clenched. This is what he had learned: hearing the sounds of the Russian armies approaching, the village's few remaining peasants had not been able to hide their joy. The German captain of the unit stationed here promptly ordered all inhabitants to leave their homes. The Nazis had driven them at bayonet point into these four cottages, nailed down the windows and doors, and set the cottages on fire.

Serov and I were still standing there cursing the only army in the world capable of such a crime, when two automobiles drove up. A door opened and out stepped a tall man with severe features. He was a major general of the cavalry. Calling to Serov, he asked what had happened. While I watched Serov explaining and the general nodding, my attention was suddenly attracted by a column of about three hundred Germans being escorted to the rear by some twenty Russian soldiers. The Germans were in the same physical condition as the other prisoners we had seen. As the column reached us, the general called to the young lieutenant at its head. Halting his men, the lieutenant ran up to the general, stretching himself stiffly in front of him. The general gave him an order, and the lieutenant had it passed on to the prisoners by one who could understand Russian. The Germans formed a double line, facing the road and the cottages from which bodies were still being carried. I sensed that something terrible was about to happen. Through the interpreter the lieutenant explained to the Germans the crime that had been committed against the defenseless civilians, either by them or their comrades. Death—no one knew in what form—hung in the air. Then the lieutenant shouted a command. The Germans began letting down their trousers.

"Sit!" shouted the lieutenant.

The temperature was twenty degrees below zero. Not one of the men hesitated to obey the order. They all sank down in the snow on their naked behinds. Serov turned a pale face to me.

"In half an hour they will all be dead," he whispered.

The general stared at the Germans seated in the snow. Then he produced a pack of cigarettes and handed it to the men standing near him. He even called Serov and gave him the rest of the pack. Then he lit one himself, looked at his watch, and turned to the lieutenant.

"Lieutenant," he said, "if any of these men are alive in half an hour, shoot him, and then go back to your unit."

★ ★ ★

UNFORTUNATELY, our victorious advance didn't last very long. On December 2 we entered Taganrog, and next day covered another twenty kilometers to the west, but that was the end. The Germans had managed to reinforce themselves and stop the Soviet push. In the north, however, the Russian counteroffensive had met with great success: Marshal Zhukov, having halted the attack led by Generals von Boch and Guderian, had begun thrusting the Germans back from Moscow, Leningrad, and Tula. By December 10 Leningrad and Moscow were safe and never threatened again.

In Taganrog I had an opportunity of witnessing the NKVD at work. It was well known that partisans had played a tremendous role in weakening the Germans behind their lines. These units consisted mainly of escaped prisoners, men who found themselves in the rear as a result of unexpected attacks, and troopers parachuted into occupied areas for "underground" activity. The majority of this last group belonged to the NKVD. Their main task was to observe the behavior of the Soviet population vis-à-vis the German occupant. They checked up on those who were friendly to the Germans, those who rendered them services, and they paid special attention to active collaborators. The result was a series of denunciations and quick trials in the recently reconquered territory. Now, there's no doubt that Russians who made things easier for the Germans deserved the severest punishment. Much advantage, however, was taken of this procedure to settle personal dislikes and hatreds. A man or woman, for instance, could be condemned for handing water to a German. For this, as well as for a case of out-and-out collaboration, the sentence was the same— the gallows. In order to make more sure of the airborne trooper's loyalty, the NKVD authorities carefully moved his family into the interior so that they would not fall into the hands of the Germans and their lives would depend on whether the man were loyal or not.

Two days after we had passed through Taganrog, Serov and I were given a three-day furlough. We decided to take a trip to the reconquered city, where some relatives of Serov had lived before the war. He did not expect to find them, but we intended to look for them just the same. The town was almost intact. In the center of the main square stood some newly constructed gallows—eight of them in a row. The public trial of Russian collaborators was already taking place. The "trials" of the fifty-odd men and women lasted no more than three minutes each. The jury was composed of a few army men, a civilian, and another man, probably the town's chief partisan. The judge was an officer of the NKVD who, during Taganrog's three weeks under German occupation, had been hiding in the city. The first eight accused were led to the gallows before the completion of the trial. These gallows were "motorized"—trucks had been placed under the ropes. The condemned were shoved onto the trucks with the help of rifle butts, the rope was tied around their necks, the trucks driven out from under them, and the eight corpses were left dangling in the air.

"I'd not like to be the driver of one of those trucks," Serov remarked.

We were about to leave when Serov suddenly grabbed my arm. Motionless, he was staring at the group of people in front of the jury. I pulled him by the sleeve. He stuck his fingernails into my elbow and seemed unable to speak. At last he turned to me.

"See that woman next to the gray-haired old man?" he whispered.

I nodded, noticing that Serov had turned pale.

"She's my aunt, my father's sister!"

I was so frightened that a minute passed before I could speak.

"Man, save her!" I let out at last. "Yell at them! She's your aunt!"

Serov looked at me as if I were insane.

"Are you mad?" he cried. "How could that help? I'd only get myself into trouble. Do you realize who you're dealing with?"

I was indignant.

"After all," I shouted nervously, "you're a front-line soldier, decorated for valor. Tell them, or I'll do so myself!"

People began staring at us. Serov laid a heavy hand on my shoulder.

"For God's sake, Fred," he said, "shut up, unless you want to lose me!"

The urgency in his voice made me keep silent. The next group of eight, which included Serov's aunt, was now being shoved up to the gallows. Pushing his way through the people in front of us, Serov stood under the trucks and stared at them. The woman seemed to sense his eyes on her and looked down. A flash of recognition crossed her face; she stood stock-still, opened her mouth, but no sound came out. Then an NKVD butted her from behind with his rifle, and she lurched forward.

I watched Serov gazing at her; he looked as dead as a statue. With a feeling of pressure on my throat, I pulled at his arm.

"Let's go," I whispered. "You're not going to watch this, are you?"

He shook his head.

"Wait!" he grunted over his shoulder.

The performance we had already witnessed was repeated. Ropes were dropped over the necks of the five men and three women on the trucks.

One of the men resisted with a shriek. Two soldiers promptly grabbed his arms. The trucks moved from under them. Eight bodies hung suspended on the gallows, their legs and arms dangling. Serov went on staring till his aunt's body, unhitched from the rope, had been dragged away by the legs and thrown, with the other corpses, onto a van. Only then did he turn to me, his face deathly white.

"Come, Fred," he whispered, "let's go to her house and find out what led to this."

We spent some time wandering around the narrow streets because Serov could not remember the exact address and he did not want to inquire about his people by their name. Finally we found the house, almost outside the town. On the lock of the garden gate we saw the stamp of the NKVD. The house was empty and sealed. Serov was standing cursing before the gate when a man in civilian clothes, but with the typical NKVD "look," appeared from behind the house. Approaching the gate, he asked us sharply what we wanted. Serov's face indicated that he wanted to kill him. He replied, however, that what we wanted was our business and asked the man about the owners of the house. The man eyed us suspiciously.

"The owner of this house," said the man, "was court-martialed for hiding a wounded German. Two of her children ran away, but they will surely be dealt with. And you?" he asked suddenly. "What have these people got to do with you?"

"They're the family of one of my friends," Serov lied quickly. "I happened to remember the address and just dropped in."

The man immediately showed more interest.

"And this friend of yours, where is he?"

Serov gave him a gloomy look.

"Killed at the front," he muttered.

The man obviously wanted to make some further inquiries, but since he was in civilian clothes and we in uniform, he no doubt realized he wouldn't get much out of us. Mumbling something under his breath, he turned and disappeared behind the building. Serov then passed along to the next house and knocked at the door. When an elderly woman opened it, Serov said:

"Anna Pavlovna was my aunt. I would like to know why she was executed."

The woman looked worried. She gestured for us to come in, but Serov shook his head.

The woman then told us the following story: In Anna Pavlovna's house, as in almost every house, three German soldiers had been billeted. One of them, a young boy, very much resembled Anna's son, who was somewhere at the front. Anna took a liking to him. By the time the Germans were about to leave Taganrog, she felt as if she were losing someone very close, someone who reminded her of her beloved son. The night they left she was roused by a knock at the door. Opening it, she was confronted by the young German. He was badly wounded in the leg and chest. In this state he could not possibly escape. Without thinking of the consequences, Anna suggested his waiting till the Russians came. As he was wounded they would do him no harm; he would be taken to hospital and be through with the war. The boy had fever and was not properly conscious of what was going on. He let her undress him and put him to bed. Next day, when Anna Pavlovna went to the local committee to report the presence of a wounded German in her house, it was too late: the Russians were already in the town. Before she reached home the NKVD had removed the wounded boy and were waiting to arrest her for "favoring the invader." On seeing the NKVD men at the gate, Anna's two teen-age children had fled. What happened then we already knew.

Serov's eyes were wild as he got into the sidecar. I didn't bother to ask him where he wanted to go. I guessed he'd seen enough of Taganrog. So I drove out of the city and headed back to Rostov.

Our divisional staff were billeted in the main street. Having found a place to sleep, we went out to a wineshop where we ran into several acquaintances, among them the young lieutenant who had been with me on the observation post during the battle of Rostov. We sat down over a bottle of sour wine and began telling one another front-line stories. The room was gray with smoke and extremely noisy. Suddenly a hush fell over it. All eyes were turned toward the entrance. In the doorway stood a girl in uniform with a sergeant's insignia on her collar. She wore a black fur cap [caps of Red Army soldiers are gray] and was accompanied by a major. When the couple had seated themselves, the babble of voices broke out again.

"Who's that?" I asked my friend the lieutenant.

The answer was a strange story. The girl's name was Marfusha, popularly known as "the black Marfusha" on account of her cap. Like thousands of Russian girls, she had volunteered for front-line service.

Unlike most of them, however, Marfusha had a fiance from whom she never separated. The boy was young, handsome, and courageous. Together they went off on patrols, deep into no man's land, even behind the German lines. The girl was a first-class sniper, with more than a few Germans to her credit. During the first day of the Rostov offensive, they had both fallen into the hands of the Nazis. The Germans, realizing that they were dealing with lovers, first raped the girl more than a dozen times under the boy's eyes, and then, bringing her back to consciousness by douching her with water, they tortured the youth to death. Thanks to an almost superhuman constitution, the girl lived through the ordeal and by some miracle managed to escape. Having wandered about the snow-covered steppe for two days and nights, she'd been picked up by Soviet sentries in a state of utter exhaustion. Back in her own regiment she had regained her physical health, but the experience had proved too much for her nervous system. She was definitely not quite sane. But it had not prevented her from continuing her career as a sniper. Working all alone now, she would still manage to penetrate behind the German lines, picking off Nazis while hiding in ditches or in the tops of trees. During the last few days she had become an almost legendary figure on our sector of the front. Some stories had her in two places at once, others that she disposed of a hundred Germans a week. In appearance Marfusha was definitely beautiful, with bright golden hair over distinctly tragic features. Her eyes were never still; she kept looking from face to face, as though still in search of her boy whom, so the lieutenant told me, she refused to believe dead.

★ ★ ★

BY December 10 I was back on the line again. The Germans were now reinforcing their positions, but the initiative still lay in the hands of the Russians. The cold was now intense. Fortunately we were doubly equipped to withstand these subzero temperatures: first, we were properly clothed, and second, we could get vodka. In fact, we received vodka regularly twice a day. It came in small glasses, corked and sealed. You pulled a string and out jumped the cork; you drank the fiery contents at one gulp, then threw the glass away. I must admit that I drank more than my share. One of my friends in the reconnaissance platoon was the sergeant major, with whose assistance I managed to put away about ten shots a day. This was definitely too many, but somehow I never got properly drunk. Only once did I fall into an orgy, and the man responsible was the lieutenant who told me the story of Marfusha.

I happened to run into him one evening in Rostov, where I had been sent to fetch some documents. He insisted that I dine with him in the mess of the unit to which he belonged. At first everything was a trifle official and stiff, and I even felt a little strange, as a noncom among officers. After supper, however, the situation radically changed. If there were thirty men in the mess, in an hour there were just as many empty bottles on the sideboard. They drank a strong *samogon* vodka which goes to the head like lightning. The evening's entertainment began by firing revolvers at the empty bottles. Then some women appeared whose profession was unmistakable. The party was growing gayer. Since I could not refuse those who wanted to drink with me, at about midnight I was very drunk, yet aware of what I was doing. It was just that my movements were not well coordinated, and I walked without equilibrium. The room looked awful, everything covered with remainders of food and broken bottles. Suddenly a young lieutenant, all his buttons undone, leapt onto the table. Kicking plates and bottles onto the floor, he shouted:

"*Tovarisch* officers, how about a cuckoo game?"

The roomful of drunks roared with delight. Those who had taken off their belts and pistols ran to the hall to get them; the scared women rushed screaming to the next room. Still retaining a vestige of common sense, I was almost through the door when a young officer, loading his *nagan*, called out:

"What's up, sergeant? Afraid?"

I must have been pretty drunk, for the very suggestion made me indignant. Taking my *nagan* out of its holster with what I hoped was a self-assured expression on my face, I waited for further developments. I'd often heard about "cuckoo," but I'd never seen it played and certainly never dreamed of taking part in it. The preparations did not last long. From the next room they brought in a desk, a few heavy armchairs, some more bottles. Then everybody filled another glass and raised it high.

"*Na zdorovye!* Your health!" roared the drunken crowd.

While they gulped down their drinks, I secretly poured mine over the floor behind me. After a warning, the light went out and immediately as many shots rang out as there were men in the room. The bullets were fired at the ceiling to indicate the beginning of the game. At once the room became silent. All you could hear in the darkness was some furniture being pushed about. This "silence" lasted about five minutes. Nobody could decide to begin the game. Finally, from the direction of the sideboard, a voice shouted the first "cu-ckoo" I felt pretty sure it was the young officer who had first suggested the game. The word was barely out of his mouth when a dozen bullets crashed into the cupboard. In answer, I heard him roar with laughter, promptly followed by the same voice yelling "cu-ckoo" from another corner of the room. Once more the revolvers opened up and again his laughter rang out. Suddenly "cu-ckoo" was shouted by someone else, this time right beside me. Instinctively I pressed myself to the floor. Most of the bullets spattered against the wall above me, but one ricocheted off it and whistled past my ear. I felt instantly sober. Remembering the sideboard—a solid, heavy piece of furniture whose lower shelves were stacked with bags of flour—I began crawling in that direction. The game was now at its height, "cu-ckoos" and bullets flashing from every corner of the room. Reaching the sideboard, I managed to squeeze myself safely behind it. Having made sure that the sacks of flour were properly protecting my body, I carefully stuck out my head and yelled "cu-ckoo." The moment I withdrew my head, the sideboard shuddered and splintered from the barrage of bullets. Some of them, missing the sacks and my body, pierced the wood and landed with a dead thud against the wall. No, I thought to myself, I will never play this game again. Hardly had I made this silent decision when another burst of shots rang out. Someone uttered a frightful scream and the room resounded with a tirade of curses. The lights went on. Behind an armchair, in a pool of blood, lay the unit's doctor, a captain, his arm shot through in two places. After vodka had been poured into the wounds, the doctor was carried into the next room where the scared prostitutes were still sitting.

    I was now almost sober and fed up with the whole thing. But again I was forced to drink. We drank heavily until I was drunker than before. Then we ran out of vodka. The whole establishment was ransacked, but not a drop of vodka could be found. One of the prostitutes suggested that the officers give her some money and she'd go out and get some "odekolon" (Eau de Cologne). A few minutes later she returned with an armful of bottles of toilet water. The water was equally distributed—half a glass per person. Now the drinking etiquette changed. Instead of gulping down one large "toast," the glasses were placed in a circle and drunk, one after the other, while the rest watched to see whether anyone attempted to get out of it. The majority, probably quite used to this type of liquor, drank quite willingly, but some began to retch and spit. Even half a glass of this abominable stuff, however, seemed to make everyone roaring drunk. It certainly did me. I find it hard to describe the effect it had on me. Not only was I utterly nauseated, I also lost complete control of myself. I was so drunk I have no idea to this day how that orgy ended. I woke up with a splitting headache and a feeling of paralysis. I looked around. I was lying on the floor among broken bottles. My lieutenant was next to me. I raised myself on my elbow. The entire party was prostrate—unconscious bodies lay on the floor, the couch, and the table. I dragged myself along the floor in search of a sink. In the next room I found the wounded doctor lying on a couch with one of the prostitutes. I reached the sink and stuck three fingers down my throat. After vomiting for half an hour, I fainted.

★ ★ ★

THE front was now unusually quiet. neither the Russian nor German line was continuous or properly defended against possible attacks. We knew that the German line opposite our sector was broken at intervals of one to two kilometers. The same thing was true of the Russian line. Realizing this, neither side could afford to take any decisive step. Both, moreover, were suffering from a shortage of human material. Front-line activity, therefore, was confined to a mutual inspection of forces. Every night patrols would penetrate the opponent's territory, inspecting the width of defense lines, reconnoitering the terrain. It was very unusual for these patrols to attack any unit or to explode enemy supplies of gasoline or ammunition. It was as though an unspoken agreement had been established, forbidding such tricks to be played.

Meanwhile I made two more applications for a transfer to the Polish Army. Shortly after the second application, I was called to the commissar. I ran through the snow to his dugout, convinced that at last I was to be transferred. My hopes were soon dashed. The commissar's face did not predict anything good. Ordering two officers to leave the hut, he waited until the door slammed behind them, then got up and looked at me. Suddenly he showered me with a stream of the most horrible curse words.

"What, you don't like our army?" he yelled. "You want to go where you have to say 'sir' to an officer? Where only a *pomieshchyk* (landlord) or a capitalist can get a commission? You want to quit a democratic, workers' army for one of those bourgeois outfits?" And again a shower of curses and insults.

There was no sense, I knew, in explaining that he was speaking of an allied army which would be fighting against Nazi Germany. I just kept silent, which made him even more furious.

"What, have you lost your tongue? You make one more application and I will personally see to it that you get your transfer, but to the White Bears!" [slang for the Siberian concentration camps]

I saluted and, without a word, walked out.

I was now definitely convinced that I could not get away by any normal, legal means. I just had to escape and that as soon as possible. The refusal of my application and the threats connected with it were an open violation of a pact between two allied governments. Such dishonesty certainly gave me a moral right to take justice into my own hands. I knew, on the other hand, that I could not desert without adequate preparation and careful thought Now that the front was stabilized, a man could no longer wander around on his own in a uniform, calmly claiming to be a survivor of some unit destroyed at the front. The work of the NKVD was far too efficient for that. What's more, I had a pretty good idea that they kept an eye on those continually applying for transfers. So I decided to wait for an opportunity and not to make any stupid mistakes.

Two changes in personnel affected my life at this point—one pleasant, the other very much the opposite. A Pole by the name of Stanislaw Krupa was assigned to my platoon. Krupa was from Stanislawáw and both his war record and his plight were much the same as mine. Like myself, he wanted to escape to the Polish Army. The other change took Serov out of my sidecar and put in his stead a fellow with an insolent face, probably an "inside" man from the *Komsomol* (Youth's Communist Party). He followed me to the point of almost assisting me in my physiological functions. I was not even allowed to ride the motorcycle alone, but had to take my new partner along wherever I went. Every two or three days we set off on patrols behind the German lines. This job was relatively simple, considering the discontinuity of the line. Sometimes, when penetrating as deep as ten kilometers behind the lines, we created tremendous joy and surprise among the peasants by unexpectedly showing ourselves. At first I didn't understand the purpose of these patrols. Sometimes we seemed to sacrifice an entire day to sneaking through the German lines simply to enter a cottage and then return. Later I discovered that each time we set off, the patrol commander brought along instructions for the partisan troops, sometimes even arms and ammunition. These patrols became so much a matter of habit that I felt nervous only once at the risks involved.

I happen to remember the date was December 21 and that overnight the recent appallingly low temperatures suddenly ceased.

As we set off at dawn the thermometer had almost reached zero. The patrol on this occasion was a large unit, consisting of infantrymen in addition to our reconnaissance platoon. The H.Q. of the sector was planning a local attack in order to shorten the line, and it was our job to reconnoiter the territory in advance. The route we were to take had been mapped out by the divisional staff. It was therefore new to me as well as to the thirty men and the three officers in command.

We had passed the first line of German trenches before daylight. Even so, we had to move very slowly on account of the snow, which was deep and so far untouched. Some three kilometers further, however, we came across a well-trodden track running at right angles to ourselves. This I found alarming. Since I was up front, I informed the commanding officer. He took it very lightly, and we continued. Krupa, marching beside me as usual, began to get nervous.

"I don't like the look of things," he kept repeating.

Nor did I. One kilometer further and we ran into another footpath. This time I tried to convince the officer that we were probably in danger, since these paths ran parallel to the front line, thus indicating that food and ammunition were being transported through here to the trenches. But the commander, a young captain, saw nothing to worry about. According to him, supplies passed down these roads only at night and just before dusk. He then ordered us to concentrate in a single formation, rather than in scattered groups. So we advanced bunched together, passing from open fields into a wood. Here we were safer, but the wood couldn't continue forever. I was getting more and more worried. So was Krupa. The wood began to thin out until it finally ended. We faced a wide-open space, beyond which another wood began.

Before taking to the open, I halted with Krupa and Serov, gazed cautiously around. The captain, catching us up, gave me a sarcastic smile.

"What's wrong?" he muttered. "Feeling scared?"

I pretended I hadn't heard. Then, in one concentrated bunch, we marched out into the open. My nerves were strained to the limit. I knew something was bound to happen. At that moment we heard the bark of a machine gun, and instantly a dozen Germans jumped from behind the trees of the opposite wood. Out of the corner of my eye I saw Serov and the captain fall. I also saw, not more than thirty meters in front of us, the spot where the machine gun was hidden. Since retreat was now useless, we all instinctively dashed forward—I with my eyes glued to the machine-gun nest. While running, I snatched a grenade from my belt and pulled the string of the safety pin. Just as I was swinging my arm another machine gun opened up on my left and I felt a sharp pain in the calves of my legs. As I fell, my hands went numb and I had a vision of the grenade dropping into the snow a few meters away. I instinctively pulled my legs up to my body and ducked my head between my shoulders. The grenade burst with a roar. Fragments whistled over my head. I felt myself losing consciousness from pain. . . . The last thing I remember was seeing a German running straight at me, waving an automatic which he held by the barrel. Would he or would he not, I wondered, hit me over the head with its butt? Vaguely, as in a fog, I saw him pass me by. Then I fainted.

When I came to, all was quiet. Judging by the sun, I must have been unconscious for three hours. My first sensation was terrible pains in both legs. I felt that both boots were full of blood. I was lying on my stomach in deep snow. Behind me, in the wood, I could hear someone groaning. I raised myself slightly on one elbow. To my left I could see four bodies—among them those of Serov and the captain. I turned over. More corpses. In all I counted eighteen. What about the rest? Where were the Germans? I turned over on my back and began to examine my wounded legs. By the holes in the felt I could see that several bullets had pierced right through both boots. I was certain the insides of the boots were wet, and that whatever happened I must stop the blood flowing. But when I tried to pull up my legs I fainted again.

When I came to, the pain in my legs had grown even worse.

Clenching my teeth, I looked for a razor in my bag and cut open the upper part of the boots. I also ripped up my pants and underwear. But I could not see the wounds properly; they were too drenched in blood. In my fur jacket I had an extra pair of clean flannel foot-wrappers. These I tied tight around my calves, praying that they would stop the flow of blood.

The wounded man behind me was still groaning. I was about to try to crawl over to him, when I heard voices and branches being broken in the wood. I promptly turned over on my stomach and lay still. The men were speaking German. My heart was beating like a hammer. If they were to discover me alive, I wouldn't stand a chance. The voices came closer, while the groans of the wounded man grew steadily louder. Suddenly a shot rang out. Instantly the groaning stopped. The Germans must have been in a hurry, for they quickly passed by me and went on. Then I fainted again.

I kept losing and regaining consciousness. When evening came I was terribly exhausted, and my wounds were hurting unbearably. By crawling a few meters at a time, I managed to reach the man who had been groaning. He was now cold and gray. They had shot him behind the ear, the way you finish off a wounded horse. I kept wondering how on earth I was going to save my life. I couldn't crawl more than a hundred meters on my own. I knew there was no village in the neighborhood. Tomorrow the Germans would be here to take the felt boots and fur coats from the corpses. I could see very little hope for myself. Just before dark I made a tremendous effort and managed to pull myself up between two old tree trunks. There, protected from the wind, I put my legs on a stump, higher than the rest of my body. But I could feel fever spreading through me, and my legs felt as though they were on fire. I had some strange visions and once more I lost consciousness.

Next time I opened my eyes, stars were shining above me. I felt somewhat better, but the fever was still high. I began chewing bacon to keep myself awake. I feared that if I fell asleep my legs might change their position and so increase the pain. I lay there gazing at the stars, at the Great Bear with its broken shaft aiming at the North Star. To kill time and stop myself thinking, I began to count the stars.

I don't know how many I'd counted or whether I'd already given up when I heard voices in the direction from which the Germans had come. I grabbed my *nagan*. If they were more Germans, I was not going to sell my life cheap. The voices drew closer and closer. Then I distinctly heard someone say in Russian:

"It must have been just about here!"

I screamed as loud as I could. The sound I made seemed little more than a whisper. I tried again. It seemed an eternity before I heard the squeaking of snow under boots and—before my eyes, blotting out the stars—saw the kind, concerned face of Krupa. He had with him a dozen men, including one of the three officers of our patrol. Having made sure that I was the only man alive, they collected all papers from the corpses and buried the dead men in a common grave of snow. Krupa then carefully hoisted me onto his shoulders and we began the long journey back.

"You've been out here thirteen hours," Krupa told me.

I didn't care how many hours I'd been out there, for at every movement Krupa made the wounds in my legs hurt beyond endurance. Fortunately I fainted again.

I regained consciousness in my platoon commander's hut. Since the regimental doctor was away, Krupa was carefully peeling the strips of flannel off my legs. Having washed out the wounds in warm water and dressed them with bandages, he wrapped my legs in a blanket and tied them well with string. I asked him what the wounds looked like. "A little black by now," said he. "But you're going straight to hospital in Rostov. They'll soon fix you up there."

"How many of the unit escaped?" I asked him.

"Eleven, counting you," he told me. But what he didn't tell me was that after I'd been hit, the survivors, led by Krupa, had managed to outwit the Germans, escape into the forest, and reach the Russian trenches by noon. Krupa had then stubbornly insisted on returning to the scene of the catastrophe to see if anyone were still alive. When one of the officers agreed, they picked a dozen volunteers and set out, this time observing every possible precaution.

"You were lucky," was all Krupa had to say as he helped carry me from the hut. Outside, they laid me on top of a truck. The truck moved off, lurched over a bump in the road. Thrown on my side, I hit my head against a metal edge and again fainted. The next thing I remember was being carried into a large building where a couple of medics laid me on a stretcher, covered me with a blanket, and left me in a huge room. The air stank of blood, pus, and disinfectant. Men on stretchers littered the floor all round me. Some of them were moaning; others were very quiet. A nurse leaned over me with a pail in one hand.

"Thirsty?" she asked.

I nodded weakly. I could barely speak, as my tongue was literally sticking to my palate. She dipped a tin bowl into the pail, raised my head, and poured "tea" between my burning lips.

"What's wrong?"

"Legs," I whispered with difficulty. She took the blanket off me and stared at the packages Krupa had made of my legs.

"When were you wounded?"

"Yesterday morning," I replied.

She nodded and disappeared through one of the numerous doors leading to the hall. After a while she returned with two medics.

"Prepare him for the operating room," she ordered.

She bent over me.

"Don't worry, young man, you'll be all right."

She looked at me again.

"You remind me of my son!" she added unexpectedly.

The medics pulled the blanket off my legs, cut the string, unwrapped the bloody bandages, and placed me on a high wheeled stretcher. They covered me with a sheet and pushed me to the end of a long line of similar stretchers. One of them put a lighted cigarette in my mouth.

"Well, brother, you're lucky," he said. "Most of you boys have to wait two days to get into the operating room. Do you know that nurse?"

"No, I don't. But she told me I remind her of her son."

I watched them smile at one another. Then one of them leaned over me.

"She lost her husband and two sons at Leningrad," he explained confidentially. "Now she is constantly looking for someone resembling them."

The line of tables moved very slowly. Near noon I reached the door to the operating room. I could not understand how I was to be operated on fully dressed. Although my pants had been cut, I still had on my uniform, the sheepskin jacket, my cap, belt, and *nagan*. I asked the medic what he thought.

"Don't worry," he said. "Keep as near your clothes as you can, otherwise you'll never find them again."

At last the door opened, and out rolled a man covered with a sheet. His arm hung awkwardly from the stretcher. One of the medics tossed it back over the body. I let out a curse. The medic laughed.

"Don't worry," he repeated. "That one won't need an arm any more."

"What . . . ?" I muttered.

"Well," he said, "the knife doesn't cure everyone, you know." My stretcher was rolled across the threshold. The room was huge. There must have been a dozen surgeons and a staff of nurses and medics at work.

The doctor they wheeled me up to was quite young. He threw the sheet back, looked at my legs, and turned to the sister:

"Please get me some new gloves. Mine are torn."

He was washing his hands with his back to me. Then he put on his new gloves and said:

"Unfortunately we haven't any ether, but you'll be given a local anesthetic. You won't feel anything."

I sat bolt upright on the stretcher.

"What are you going to do to my legs, *tovarisch* doctor?" I asked.

He looked at me in surprise.

"Well," he said, "the right leg has to be amputated under the knee. It's black all over from gangrene; the left one I can leave on for the time being, but it should probably also come off." Instantly all my energy returned. I refused to believe amputation was necessary. He had hardly looked at my wounds.

"I am against amputation," I said calmly. "Please clean my wounds and leave me alone. Every wound I've ever had has healed like a dog's."

The doctor looked annoyed.

"Nobody asked you whether you want the operation or not. This is a front-line hospital, not a nursing home. We have to amputate. And if we have to, we will."

I must have given him a hell of a look, for when I said, "You will amputate nothing!" he stepped back with a startled face. He glanced at me for a moment and then turned to the sister:

"Ask the chief surgeon to come here."

He waited a little while.

"Where are you from?" he asked.

"From Cracow."

"Where's that?" His knowledge of geography, apparently, was confined to the USSR.

"In Poland."

He looked at me with interest.

"Ah! You're Polish? But all the Poles were withdrawn from frontline units."

"As you see, not all," said I.

Then an elderly man appeared, accompanied by two assistants. My surgeon explained the situation. The older man bent over me.

"When you're told that an operation is necessary," he roared, "it is necessary! Understood? What's more, don't you forget—your life belongs to the state, and in hospital the state is represented by the doctor!"

This was too much.

"My life belongs to me, *tovarisch* doctor," said I. "I lent it to the state for the period of front-line service. But whether the state can amputate my leg is a question I shall decide."

They all stood there speechless, staring at me as at an idiot. Then the elderly man exploded:

"You probably have high fever. That's the only explanation for your lack of discipline." He turned to the young surgeon. "Operate!" he ordered.

"You can operate on your own arse!" I roared, "but don't you touch me!"

Now the reaction was different. The chief surgeon howled like a wounded bull:

"Slap him down and operate!" With one movement I grabbed my *nagan* and aimed it at the yelling surgeon.

"*Tovarisch* chief doctor, if you want to operate on me, you'd better order your own coffin first!"

He stood stock-still, pale and speechless. No one spoke. They seemed hypnotized by the barrel of the *nagan*. Finally the young doctor looked at me and asked calmly:

"Will you assume the consequences if there is no operation?"

"Naturally," said I.

I felt myself growing weaker and the fever mounting. I kept praying that I wouldn't faint. If I did, I knew they'd cut both my legs off, if only to keep their prestige.

"I'm sure I'll pull through," I added.

Although he had not been given any order, the young doctor began dressing my wounds. He washed them, extracted some bone fragments from the right calf, and bandaged my legs. The rest stood by in silence, watching. I was still aiming my *nagan* at the chief surgeon. As the younger surgeon was tying the last bandage, he whispered in my ear:

"It's possible that these legs can be saved. It all depends on your constitution."

He spread the sheet over me and turned to the medics:

"Take him to Section Eleven and tell the sister on duty to give him the bed the young *politruk* vacated today. Don't forget!"

They wheeled me out of the door headfirst, through corridors packed with soldiers. Vaguely I felt myself being lifted onto a bed and my clothes being taken off. I must still have been holding onto the *nagan*, for the last words I heard were:

"Leave it in his hand, but unload it. We don't want that kind of accident here."

★ ★ ★

ABOVE me I saw a dirty, cracked ceiling, with a large cobweb in one corner. I felt too lazy to turn my head and see what was going on around me. The light was hurting my eyes, so I kept them half-closed. On the ceiling I discovered some interesting things. The few spots caused by water leakage had the strangest shapes. The largest blemish had the contours of Africa. I followed the continent's line from Tunis, down the Ivory Coast, all the way to Capetown. Even the Island of Madagascar was there. When I closed my eyes and opened them again the map had changed to the head of a wolf. I preferred Africa. Tracing the "coast" with my eyes, I remembered a geography test in which I had been flunked by the professor for not being able to find Lake Victoria. Now I found the spot with ease. I went on, tracing the Nile, the ranges of the Atlas Mountains, I was warm and comfortable. Yet I was so overcome with a feeling of weakness that when I heard the sound of a door opening I could not summon the energy to move my head. So I lay there waiting until I saw—covering the image of Africa—a female face surrounded by a white scarf. She smiled and muttered something I couldn't hear. I closed my eyes to indicate that I did not want to look at her. I wanted to tell her so, but I felt too lazy to open my mouth. When at last I did open my eyes, she was still there, and still smiling. She spoke again. This time I understood her. She was asking how I felt. Only then did I realize there must be a reason for me to feel bad.

She was quite young, with a red cross on her white cap. So she was a nurse! As it began to dawn on me where I was, I was aware of pain in my legs. Now I knew that my unwillingness to move my head was a result not of laziness, but of weakness. Though the girl was still speaking, I wasn't listening. Now I remembered everything. The scene in the operating room, the fight with the chief surgeon. A terrible thought sprang to my mind: had they amputated my legs? I did feel pain in my calves, but this did not reassure me. The nurse's face looked slightly alarmed. She had probably seen fear in mine. I stretched out my hand, down along my body. I moved it slowly, partly on account of the effort, but also from fear that my body under the blanket would suddenly end. My hand had reached my thigh, then one knee, but it could get no further unless I sat up. At last the nurse seemed to understand what I wanted. Still smiling, she lifted me carefully into a sitting position. I stretched my hand out as far as the calves just above the ankle, where I felt a thick dressing under my fingers. The legs were there! With a feeling of intense relief I closed my eyes. The nervous strain poured out through all my body. I was again exhausted. The nurse sat me up against the pillows and began stroking my hair.

"You have your legs," she whispered. "Don't worry. If you hold out until this evening you will live and walk again."

Now it was my turn to smile.

"How long have I been unconscious?" I asked.

"Two days and nights," she smiled. "Today is December 24 and they brought you in here the morning of the twenty-second."

She began fixing my pillows.

"At first your fever was so high we feared you'd die. But thanks to your constitution and my brother's care, the real danger is over."

I sighed with relief.

"But what about the gangrene?" I asked her.

"That was not gangrene," she told me. "Just blood clots under the skin, made worse by frostbite and the remains of your avitaminosis condition. Your legs are nowhere near so black today."

Then she left me.

So I had saved my own legs! The thought of it brought beads of sweat onto my forehead.

After a while she came back, carrying a plate and an enamel cup.

"You've got to eat something," she said, holding the cup to my mouth.

Only now did I realize how hungry I was. Watched by the smiling nurse, I devoured two plates of goulash and drank two cups of tea. I was just finishing off the second plate when the door opened and a young man in a white smock came in. His face seemed familiar. He came up to the bed and looked at me, smiling.

"I see you feel quite well," he remarked. "I'm glad."

"I know your face, *tovarisch* doctor," said I, wondering where I'd seen it before.

He smiled broadly.

"You should," he said. "I'm the man who nearly amputated your legs!"

Suddenly my old anger returned.

"But for my common sense," I found myself shouting, "you'd have ruined my life!"

He stopped smiling and sat down at the foot of my bed.

"Young man," he began, "you must not blame me. Four months I have been working in front-line hospitals. Every day, sixteen hours a day. Always in a frantic hurry, because wounded men need help in a hurry. Five times the number of doctors we have would not be enough. We have no time to think. When a wounded man's life is in danger, we operate. To attempt a cure, we have neither the time, the space, nor the staff. So try not to be surprised at what happened to you. Judging that the gangrene was moving up the body, I decided to amputate. I had not the time to take the tests necessary to discover whether the discoloring of the legs was caused by gangrene or by frostbitten blood vessels."

While listening to this explanation, I felt my indignation slowly receding. This man was telling the truth. He had kind, frank eyes which flickered with fatigue. I raised myself on the pillows.

"*Tovarisch* doctor," I said, "please forgive my reproaches. I had no idea that the problem of medical personnel in hospitals is so sad. I thought such sudden decisions to operate were prompted by a lack of interest, that the human body was approached as a butcher approaches a hunk of meat. After what you say, I understand and realize it is almost impossible to avoid mistakes."

He smiled at me again,

"Good," he said. "Now that you understand, let us forget what could have happened. Let us be happy that it didn't and that you'll soon be able to walk again!"

The nurse had been listening to our conversation in silence. Now she came closer.

"If you could only see, Volodia, what an appetite he has!" she said to the doctor. "I'm sure his legs will heal sooner than you think."

I was surprised by the informal tone in which she spoke to the doctor. The latter caught my puzzled expression and explained: "This is my sister, sergeant. Her name is Masha."

Now I remembered that the girl had mentioned her brother who had taken such good care of me.

"I want to thank you, *tovarisch* doctor," said I, "for having looked after me so well during those first critical hours. Your sister told me that my recovery is due chiefly to you."

He smiled. "That makes me feel better," he said. "First I wanted to amputate your leg, then I tried to make up for it by watching over your life. You might as well know that for two days it was touch and go."

When I woke there was a kerosene lamp on my night table, and beside it a pine tree branch, a piece of sausage, and two hard-boiled eggs. I was staring at these objects when Volodia came in with his sister. They sat down by my bed. Volodia seemed embarrassed. I realized he wanted to say something, but did not know how to begin. Finally Masha broke the ice:

"Listen, not long ago we had a Pole from some labor battalion here and he told us that your Christmas begins on December twenty-fourth. Neither of us, you see, is religious, but our mother was, so we know what Christmas can mean. You are alone, far from home. Try to imagine that today we are members of your family."

I was very touched. But I didn't know what to say. Fortunately, and most unexpectedly, Volodia saved the situation by opening a bottle of vodka. He filled three large glasses.

"*Na zdorovye!* To better days!" he said, looking at me.

I raised my glass.

"To your success, and thank you for everything!"

We drank and ate in silence. Then we began to talk. The brother and sister were from Moscow. Their father held an important position in the commissariat of the food industry. Although Volodia had been a party member for a year, his idea of the postwar world was not at all "red." When I showed my surprise at this, his answer was brief.

"Communism is undoubtedly a wonderful idea," he said, "but it is still in the Utopian stage. We ourselves are now suffering from the fact that we considered it a system that could be put into practice immediately. I see for this world only one road to true communism—a slow, long-lasting evolution in that direction. Any attempt by us to graft it onto another country as a ready-made system will achieve nothing."

"True communism?" I repeated. "What do you mean by that?" He looked at me seriously. Then, his face breaking into a smile of rare charm, he said, "A community in which you, Masha, and I will work according to our capacity and receive according to our needs."

★ ★ ★

DURING the next few days I rapidly regained strength, and my wounds, under the daily dressings, showed such a marked improvement that my legs were placed in plaster casts.

But what casts! From the ankles over the knees! "A proper cure," Volodia warned me, "depends on keeping the legs still."

I had by now grown so intimate with Volodia and his sister that I decided to take them into my confidence and tell them that since it was impossible to get my transfer legally, I was determined to desert. They didn't seem in the least surprised. They listened, in fact, with sympathy and began to analyze with me the possibilities for such an undertaking. Volodia said he would first of all make sure that a transfer was legally impossible; if it were, then he would try to get me a transfer to a hospital farther from the front—whence, according to him, it would be easier to disappear. Meanwhile the two of them looked after me as though I were their brother.

Early on January 10 Volodia dropped in on me in great excitement.

"Nothing can be done legally," he said. "I've just had a stormy discussion on the subject with the hospital commissar. Your only hope is to get transferred to a hospital as far as possible from here. From there you'd stand a better chance. This place is far too well watched."

Promising that he'd try to get me the transfer, he left.

The same evening he was in again, very upset.

"The commissar is already on the warpath!" he whispered. "This morning I spoke about you as 'one of the wounded' without mentioning your name. Now he's suspicious. You cannot afford to wait. You must leave immediately."

Things now began to move fast. The following morning I boarded a hospital train carrying a small transport of wounded soldiers, most of them with their legs in casts. Masha, who stayed with me till the train pulled out, had discovered that the transport was headed for a hospital in N— a small town halfway between Rostov and Stalingrad. She had written a note recommending me to an acquaintance of hers, Viera W— who was employed there.

I separated from Volodia and Masha with real sorrow. During these two weeks we had grown very fond of one another.

On reaching N— next morning, we were driven on sleds to the hospital, which was in a school building. While waiting on stretchers in the hall, I asked a passing medic if he knew of a nurse by the name of Viera W— . When he said, "Yes, she works here," I asked him to inform her that an acquaintance wished to see her. I did not have to wait long. Among the stretchers a tall, graceful figure came in. She had superb chestnut hair, huge black eyes, beautifully framed and deeply set. She came up to me, looked at me carefully, and said in a tone of surprise:

"I do not remember that we ever met."

I stared at her, speechless with admiration. She made a charming grimace of impatience.

"Well," she said, "maybe you will tell me how you know my name."

When I had recovered from the first shock, I told her that Masha had spoken of her, and that she had the most superb eyes I had ever seen. Then I gave her Masha's note. When she'd read it, she glanced at me and said:

"I will try to put you in a single room in my section."

She succeeded. I could have asked for nothing better. The hospital was small and quite well organized, the food plentiful and tasty. Although there was a general shortage of doctors, it didn't affect me, as my wounds were healing under the plaster cast which was not to be changed for a month. As for Viera, she was wonderful. She was constantly in and out of my room. She would drop in to fix my pillows, to bring me a cigarette or a newspaper. One evening she told me she had plenty of time, as a new nurse had been assigned to her section. She smiled when she saw my beard, and offered to shave it. I did not want to let her, but she insisted. Having soaped my face, she began shaving me very delicately. I just sat on the bed motionless, abandoning myself with delight to these operations. She put her fingers on my lips to stretch the skin. I could not take it any more, and began kissing her hand. She stopped shaving me, laid the razor on the table, and stood there, leaving her hand for me to kiss. Though never timid, I didn't dare embrace her. Then she slowly sat down on the bed, threw her arms around my neck, and looked into my eyes. A moment later she was pressing herself against me and kissing me on the lips. I was so surprised I didn't even kiss her back.

"Aren't you going to tell me you love me?" she whispered. Having recovered from the shock, I put my arms around her. Then she leaned over and put out the light on the night table. In the darkness I heard the sound of shoes falling on the floor, and she slipped under my blanket.

★ ★ ★

WITH Viera's help, my days in the hospital passed as in a dream. We read Pushkin and Dostoevsky together; we also discussed communism. Viera believed that this war would give the Russian people the real freedom for which they had fought during the October Revolution, and of which the Stalin regime had since deprived them. Sure of her good faith, I told her of my plans for desertion.

"That's very understandable," she said. "I see no reason why you shouldn't fight the Germans among your own people."

But for the moment, she added, there was no point in making plans, as I was not in a condition to move. When the time came she would see what she could do to help me.

One morning, after my legs had been in a cast for a month, Viera wheeled me into the operating room to have it changed. In the doorway I passed a stretcher coming out. On it lay a man with his leg in a cast. I was gazing at him thoughtlessly, when suddenly I recognized the face of—Aram! Sitting bolt upright, I shouted: "Aram, Aram, it's me—Fred Virski!"

As the door closed behind me, I thought I heard an inarticulate cry from my old friend. While the doctors were changing my cast, I was so excited I did not even pay attention to the condition of my wounds. I was feverishly telling Viera who Aram was and what he meant to me. Then I looked up to find an oldish doctor, a man with a mass of gray hair, smiling at me.

"Sergeant," he was saying, "your legs will be the same as they were before you were wounded."

Viera, with no sign of embarrassment, began kissing me.

"See, Fred," she whispered happily, "you're going to lead a normal life."

I saw the old doctor staring at us from under his mop of hair.

"Love, love, always the same!" he muttered. "There's nothing new in the world!"

Viera wheeled me out of the room. Crossing the threshold, I heard a commotion in the corridor. A wheeled stretcher was surrounded by some medics who were trying to push it on. Then I saw that the wounded man had stuck his crutch between the spokes of the wheel. As my table approached the other, I recognized Aram's black, curly head. Realizing the situation, Viera pushed the medics away and my stretcher up to Aram's. Throwing his arms around me, he began sobbing like a child. Even I could not hold back my tears. I asked Viera if it would be possible to put another bed in my room. When she hesitated, I quickly added that I had no secrets with Aram.

An hour later Aram, lying in bed opposite me, was telling me his story. Six months ago, when I had left him on the Dniester line, the regiment had retreated to Odessa with tremendous losses. He had remained there during the city's siege; then, in the middle of October, they had been evacuated by sea, landing at the port of Tuapse, in the Caucasus. By this time Aram had had enough. Despite great risk, he had taken a "furlough" on his own to his native village near Yeryvan. After a few days he had returned for a reorganization of the depleted Odessan units in Krasnodar. Since there had been no checkup on the survivors of Odessa, his "furlough" remained unnoticed. Then, having taken part in the Rostov offensive, he reached Taganrog and at the beginning of January had been wounded. Fortunately, he had not been threatened with amputation, as his wounds were clean and uncomplicated. He had arrived in N— three days ago. In a whisper, Aram told me he was pretending that his right leg was paralyzed. If he could manage to convince the medical commission of this "fact" he would be given at least six months furlough to go home and rest.

Aram was a specialist in hospital matters. Having listened to the report of my wounds, he concluded that I too would probably have the right to a six months furlough because the bones had been fractured. With typical ardor, he began describing how wonderful it would be to go off on a furlough together to his beloved Armenia. The mere thought of our having nothing to do but lie around together in delightful laziness brought a celestial smile to his pale face.

I then told Aram about my plans. Though grieved that our roads would part again, he believed that for me desertion would be the wisest thing. On certain conditions. He himself had considered the possibilities of flight, but found them very unfavorable. He maintained that since the December offensive, the territorial gains and the present lull at the front, particularly strict discipline was being kept in the rear. The NKVD, which, during the Russian retreat, watched the line indirectly only by pursuing deserters, now spread its activities over the whole country. On every railroad track they employed special mobile patrols to control documents of all soldiers whether alone or in small groups. He had heard of cases of exaggerated zeal where men on furlough or returning from hospitals, having lost their documents and unable to explain themselves sufficiently, had been considered deserters and shot. According to Aram, a flight without papers covering my movements was doomed to failure.

★ ★ ★

JANUARY was coming to an end. My legs were healing nicely, but I still could hardly move because of the plaster which immobilized my knees. Viera got me a pair of crutches so that I might practice walking in the corridors. It wasn't easy. Not only was the plaster very heavy, but every time I put any weight on my feet the pain in my legs was almost unbearable. Aram also dragged himself along with me, stubbornly leaning on his left leg while dragging his right behind him.

Viera was now afraid that she might be transferred to a hospital in Stalingrad, where they needed qualified nurses of her type. For this reason she was all the more feverishly, and so far fruitlessly, looking for an opportunity to get me away. She also feared that if and when she were gone I would commit some blunder and get caught. It was characteristic of her upbringing in this strange world that she accepted our separation as inevitable.

"You belong to another world," she said. "Mine is here. This is my place. There is no sense in thinking that we might meet again. I knew it from the very beginning. That's why I have resisted any serious feeling toward you. But don't worry, Fred!"

Well, unlike her, I did not believe that this war would change the face of Russia. Nor did I know down what paths destiny would lead me. But of one thing I was sure: if I lived through this chaos, I would not remain in Russia. It was difficult for me to contemplate separation from this girl. I had a deep feeling for her and loved being with her. Little did I realize at that moment how short our time was to be.

During the evening of February 1 she dashed into our room in great excitement. She had just come from the hospital office where an officer was preparing some travel documents for a soldier going to Kuibyshev. Now Kuibyshev, Viera knew, was only a hundred kilometers from Buzuluk, where the staff of the Polish Army was stationed. She had taken a chance and asked the officer if he would like to make out another such set of documents, without the name of the soldier. The officer had at first agreed on condition she went to bed with him. When she refused he became very angry and threatened to tell the hospital commander what she had asked for, and they would find out for whom she wanted the documents. Viera, wringing her fingers in despair, kept repeating:

"Why did I refuse him? Why did I refuse him?"

I was indignant at the mere thought

"What a question!" I snorted.

She looked at me in surprise. "You don't mean to say you'd be jealous!" she said. "That would be ridiculous. We're going to separate shortly, anyway. No, I refused because I was afraid that in the morning he would laugh at me and refuse to keep his promise!"

"I am glad it didn't happen," said Aram. "If you had accepted his price you might have arranged the flight for Fred, but I bet he'd have cheated you in the end."

There was no sense in discussing it. Now the question remained whether this rascal would stand by his threat and start investigations to find out for whom Viera wanted the document.

In the morning Viera woke me at about ten o'clock. I could see by her face that something very important had happened. Having closed the door carefully behind her, she told Aram to sit by the door and signal if he heard anyone coming. Then out of her blouse she produced a handful of papers which she gave me without any explanation. I unfolded the first one. It was a temporary identification card for someone by the name of Ivan Mihailovitch Bierdieyev, "issued in place of the original, lost at the front." The photograph pasted in the corner showed a young, stupid face.

"What does this mean?" I asked.

"It looks like some strange portent," she said. "Bierdieyev is the man for whom they were writing out documents for a trip to Kuibyshev. Those documents are in your hand."

"And where is he?"

"Ah," she said, in a half-mysterious, half-excited tone of voice. "Believe it or not, Bierdieyev died this morning. His stomach and chest wounds were completely healed. He had been in the hospital six months. They think he died of a heart attack."

I felt suddenly very nervous. I realized what an extraordinary opportunity this was. At the same time my head felt dizzy at the thought of all the risks. They would probably search Bierdieyev's corpse and, not finding the documents, the suspicion would fall on Viera. What had she to say to that?

"Don't worry," Viera assured me. "Bierdieyev was buried an hour ago. The papers found on him were taken to the office. Nobody can possibly suspect that I have these documents. He was even in a different section."

I observed her carefully.

"Viera, are you sure what you are saying is true?"

"Of course," she insisted impatiently. "Don't worry about me. All we have to do is paste in your photograph and insert the missing piece of stamp. I'll bring you your clothes and you must go. Now or never!"

Aram was also hurrying me. He pasted in my photograph, penciled the section of the stamp on the picture. Now I had two identification cards. I looked at the other documents Viera had brought: a hospital discharge, a railroad ticket to Kuibyshev, a card permitting bearer to receive food during the trip, and an assignment card to the aircraft training center in Kuibyshev, where the dead Bierdieyev was to have become a machine-gun instructor. All these papers were in perfect order, complete with stamps. When Viera returned with my uniform, I dressed quickly while she prepared some food and put it in my bag.

"Fred," she suddenly said, "something has just occurred to me: Bierdieyev was to have left the hospital only the day after tomorrow. These papers are not valid until then. You'd better go to the *voyenkomat* and make one more try with your own papers. If you can get a legal transfer, burn the others. If you can't, don't, whatever happens, come back here. Hide somewhere in the city for two days, and then go."

"Yes," said Aram, nodding, "that's good advice. But don't lose time. You must get out of here."

Viera helped me into a pair of new felt boots she had found for me. They were wide and soft and enclosed the plaster casts.

The farewell was brief. Aram's face was pale and set. "Think sometimes," he begged, "of your old friend from distant Armenia."

"You don't have to ask me that, Aram," said I. "Real friends are too rare."

Viera and I kissed without speaking. Then she freed herself, turned her back to the door. "Go, Fred, before I break into tears. Please go!"

There was such insistence in her voice that I pushed open the door, hobbled down the corridor and out into the street by the back entrance.

★ ★ ★

I DRAGGED myself along the snow-covered streets toward the voyenkomat, which, according to Viera, was not far from the main square. Presently a passing sled drew up beside me. A peasant in an enormous fur cap called out:

"Where are you going, son? Want a lift?"

I boarded the sled gratefully. The peasant drove me as far as the *voyenkomat*. Leaving my bag with a guard, I hobbled toward the commander's door. A line of people, mostly women, were formed up in front of it. When they saw my crutches they encouraged me to pass through, even to go in without waiting. This sympathetic attitude toward a wounded soldier was to help me immensely during the days ahead. On reaching the door I saw a sign: "Kombryg Viltchkov."

I knocked and walked in without waiting for an answer. Behind the desk sat an elderly man with a surprised face.

"What do you want, sergeant?" he asked.

I repeated the petition that I had recited so many times. He seemed interested.

"And where are you from?"

"From the hospital."

"Discharged?"

The word "yes" was on the tip of my tongue when I remembered. "No," I said, "I am still under treatment."

He seemed to be watching me with interest.

"Why don't you put in your petition through the hospital commander?" he asked.

I told him I had done so, but the commander was unable to give me any information. He then asked for documents to prove that I was a Pole. I handed him my own army pay book. He looked it over carefully and handed it back.

"Sergeant, you surely have the right to ask for a transfer to the Polish Army, but I cannot grant it. The decision rests with the supreme command of our front sector."

I knew that this was not true, and that everything depended on him. So I bluntly told him so. Giving me a vicious look, he got up.

"*Von!* Get out!" he roared. "You're not going to teach me my duties!"

I did not dare to anger him further. If I were arrested it would be discovered that I was running away from the hospital. It would cost me a bullet in the head. It could mean the same for Viera.

I shrugged and staggered out. I sat down on my bag and rolled a cigarette. I say "sat." Actually, I could do one of two things, depending on the facilities: I could either sit very low with my legs stuck out in front of me, or very high so that my legs hung stiffly to the floor. I was about to stamp out my cigarette and go, when the guard, who was chewing melon pips, handed me his newspaper. The paper was old. Suddenly a headline electrified me:

"*General Sikorski, Commander-in-Chief of the Polish Armed Forces, visiting Marshal Stalin at the Kremlin. All Poles residing in the USSR will be inducted into the Polish Army in the USSR.*"

I feverishly read the few lines under this headline. General Sikorski had been ceremoniously received by Stalin. It was decided that the Polish Army would be transferred in the near future to southern Siberia, to the regions of Tashkent, Samarkand, and Ashkhabad. The rest of the column had been torn off. With the help of the guard I got up and with a joyful face went straight back to the *kombryg*. Pushing the paper under his nose, I pointed to the headline.

He read it, puffed up his lips.

"So what?" he asked.

I looked at him in surprise. I considered this sufficient proof that my claims were justified. He stared back at me with a malicious grin.

"How long have you been in the Soviet Union?"

"Two and a half years," I replied, wondering why he asked.

"And after all that time you don't know that what the papers say is always true, but that you can never refer to it?"

I made a surprised face.

"But this was said by *tovarisch* Stalin," I replied, desperately playing my last card.

"Don't you dare threaten me with the name of the Leader!" he yelled. "I am entitled to refer to what Marshal Stalin said, not you! Get out of here and don't let me see you again or I will have you thrown into jail! *Von!*"

I turned and hobbled out into the corridor. There was no sense in arguing with that man. It could have a sad ending. I lowered myself onto an old box and began wondering what I should do next. On a door opposite I saw a sign: "Cpt. Tchubukov, Assistant to the Commander." A captain? It's always easier to talk to a captain.

I limped over to the door and knocked.

"Come in!"

I pushed the door. Behind a desk sat a young man, warming his legs at an iron stove. Seeing my crutches, he quickly got up and helped me sit down in a chair.

"They let you out of hospital in that state?" he asked. "Those rascals simply don't care. They just want to get rid of one poor devil to make room for another."

The captain began talking about himself. He was from the infantry. Wounded at Stalino. After leaving hospital he'd been assigned to the *voyenkomat*. Then I told him about my petition.

"But of course," he said, to my amazement. "I will prepare all your papers, then all you have to do is take them over to the commander for a signature."

He sat down at the desk.

"Just the other day we had a Pole in here, but a flyer. The *kombryg* was not here then, so I did everything myself and sent him out to Buzuluk,"

Pity the *kombryg* is here today, I thought. If I told the captain I'd seen the *kombryg* twice already, he'd immediately stop filling out the documents. On the other hand, if I approached the *kombryg* a third time he'd certainly have me arrested. I was wondering what good these documents were going to be to me when the captain interrupted my thoughts.

"Here you are," he was saying, handing me some papers. "Here you have a transfer as a sergeant of the Red Army for the disposal of the staff of the Polish Army, your military railroad ticket, and this card which entitles you to receive food."

It was the same set as I had in my pocket issued in the name of Bierdieyev. The captain thought for a while, took out of a drawer two stamps, an oval one and a triangular one, and banged them on each document.

"Now off you go to the *kombryg*," said the captain. "He will add the most important stamp, a round one, sign his name, and that's all. *Bon voyage!*" He shook my hand cordially, and I hobbled away.

Positive that the *kombryg* would have me arrested or even shot, I limped over to a post-office on the other side of the street. I went up to a writing desk, and without giving it too much thought, signed each document: "Viltchkov, *kombryg*" I was in the midst of blotting the ink by sticking the papers to the plaster wall, when in walked three soldiers with bands on their arms and guns at their belts. I slowly folded the papers and stood still. My mind was working feverishly. I could not show Bierdieyev's papers, which were legal, but which would be valid only two days hence. Would these men notice that the papers I had just signed lacked the round stamp of the commander of the *voyenkomat*?

They came up to me. The younger sergeant saluted and in an official tone asked for my papers. With a bored gesture I produced the documents I had just put in my pockets and handed them to the sergeant. I watched him carefully and with a beating heart.

He had an intelligent face. But I noticed that he was running through the documents carelessly, paying no attention to the stamps. He handed them back to me.

"Filled out today?" he asked.

"Ten minutes ago," said I indifferently.

He saluted, turned, then stopped in the doorway.

"Unfortunately," he called back, "there's no train to Stalingrad today. In fact, we don't know when there will be one, as there's no coal in the whole region."

I waved my hand.

"That's all right!" said I. "I'm in no hurry."

Only now did I feel beads of sweat coming out on my forehead. I put the documents in a separate pocket, buttoned my fur collar, and limped out onto the street. My crutches either stuck in the snow or slipped on the icy surface. Crutches in snow are no pleasure. Several times I almost fell. Very tired, I wandered in the direction of the railroad station, as I did not want to take the patrol's information for granted. I could not return to the hospital. I dared not risk it, though I would have given a lot to see Viera and Aram once more. I dragged myself slowly, stopping frequently to rest.

On the empty, snowy sidewalk, a man in uniform came walking toward me. His long, well-cut coat had no insignia, so that I could not see his rank. Suddenly the man roared in a wild voice and threw himself at me with such force that he almost knocked me over. Taking me in his arms, he howled in Polish:

"Fred, Fred, is it really you?"

I was so busy trying to keep my equilibrium on the crutches that half a minute must have passed before I recognized, sticking out from under a fur cap pulled down over his ears, Ludwik's big red nose! It was really Ludwik—Ludwik, plumper than ever, very cleanly shaved, in an elegant coat and superb *valonki*. When I left him in Podgorodne he had looked miserable, dirty, in a grayish, worn uniform. Now he looked like a different man. The first questions he asked me were what had happened to my legs and whether it would not be a good idea to go immediately to his house.

Of course I accepted. He then asked me in what capacity I was now in N—. To cut a long story short I just said:

"Until this morning in the capacity of a wounded soldier, now in that of a deserter."

He was not surprised.

"Always better," he remarked, "than in the capacity of, say, a corpse!"

I followed him to a corner where a small sled, drawn by two horses, was standing. At the horses' heads I noticed a soldier who, on seeing Ludwik, quickly ran up to the sled, pulled away the fur rug covering the seat, and helped us both onto the sled. I stared at this in great surprise.

"So you're a big shot!" said I.

Ludwik lowered his voice:

"Don't talk to me in Polish in front of my people. Nobody knows I'm Polish. I'll explain everything later."

The coachman whipped the horses and away we went. We didn't talk. Ludwik simply stared at me as if I had just returned from the kingdom of death. Presently we stopped in front of a two-story house with a small garden. The soldier again helped us off, and took my bag. The door opened and an elderly woman with a noble expression stood on the threshold.

"Back so soon, Aleksiey Mihailovitch?" she greeted Ludwik with surprise.

I was even more surprised at the new names Ludwik had adopted—names which fitted him like a hunchback to a wall. We walked into the hall. Ludwik ceremoniously introduced me to the lady without mentioning a name.

"My very good friend!" said he.

I bowed. The lady disappeared, and we entered a large room which turned out to be Ludwik's bedroom. The soldier put my bag in a corner, saluted, and left. Ludwik then helped me undress, laid me comfortably on a couch, threw some wood into a tile stove, and settled himself comfortably in an armchair.

"Start talking, Fred!" he said.

Briefly, without going into details, I described my adventures. At the end I showed him my papers, those in my name and those in that of Bierdieyev. He had listened to everything with great attention. Now he looked over the documents and nodded.

"Before I tell you what I think of your plans," he said, "perhaps I'd better tell you something about myself." And this, in short, was Ludwik's story:

After Podgorodne, he had wandered about in various labor battalions until he landed up in command of an engineer corps, under whose orders all the labor battalions went. He entered the supply service as a man without a past (just as I had once), pretending he had lost his memory. But he was very consistent about this, never mentioning that he was a Pole. For a time he was head of an army storage house; then, thanks to his great talent for organization, he had risen higher and higher until now he was holding a responsible post as chief of food supply to the first line of the entire engineer corps. Though he had been given the rank of major, he seldom chose to wear the insignia. Normally this would not have been allowed, but on account of his generally strange behavior, which the authorities attributed to the effect of amnesia, he got away with it.

"So I reached this high position and with it acquired a comfortable and secure life," Ludwik concluded. "But even this life has its dark sides," he added. "You see, Fred, even if I wished, I could not do what you are doing. For me the road of retreat is closed. I cannot move around as easily as you. I have to follow the road I chose. But I don't know which of us has made the better choice. We've had enough war. If I manage to keep this job I'll never see the front again. And you know that war, observed from this distance, is not bad. But you, going to the Polish Army—even after you've broken through every obstacle and got there safely—you will start from the beginning. They will not respect your rank of a sergeant in the Red Army. Also, you are only twenty-two. You will again find yourself in a frontline unit, with a gun in your hand. I would not like to go through that again."

This was all right for Ludwik, but my general attitude to these questions was different from his. We did not discuss the matter any further.

He then described the background of his landlady, whom I had met in the hall. The family had belonged to the aristocracy before the Revolution. By some miracle, no doubt enhanced by the popularity they enjoyed among the people of all classes, they had managed to avoid the repressions, arrests, and deportations. When war broke out, however, her husband, already a senile and helpless old man, was arrested for "favoring the enemy." Having "formed his opinion" of the Germans on two years of pro-German propaganda, he hadn't been able to change it fast enough after the outbreak of war. In July 1941, he was unfortunate enough to announce in public that the Germans were decent people. Fortunately he had died a natural death before a worse fate could befall him.

The widow and another elderly woman, a cousin of hers, were informed that no one was to know of my existence. The two ladies needed no further explanation. They showed me the entrance to the cellar and a spot where, among some barrels of cabbage, I should hide in case of emergency. Ludwik insisted that they were alarmists. An hour later he was proved wrong. The landlady came into Ludwik's room with the news that a patrol of five soldiers and an officer were combing one house after the other, looking for a deserter. I hobbled down to the cellar. Ludwik put me between the barrels, carefully hiding the crutches. The patrol made a thorough search of the house. They even came into the cellar, poked about in its corners, and left. When I emerged from my hiding place, Ludwik told me that the officer had informed him that this morning a Polish boy had escaped from the hospital, and that because he was on crutches and there had been no train leaving here, he must still be in the city. I had not imagined that news of my desertion could leak out so soon. I was afraid for Viera. Ludwik suspected, no doubt rightly, that the *kombryg* had telephoned the hospital commander, bawling him out for sending a man who hadn't even got his discharge from there. Ludwik was worried too. He promised to go to the hospital the following morning and find out what was happening to Viera.

★ 199

* * *

I SLEPT badly. all night long I was tortured by nightmares. A huge dog with the face of the kombryg kept pursuing me down a narrow pathway, high above sea level. The path ended at the top of a cliff, below which lay the ocean. I turned round to attack the dog—kombryg. He sprang at me and sunk his teeth into my arm, while I lashed out at his hairy head. The situation was becoming most precarious when I woke to find Ludwik's smiling face above me. Holding my arm with one hand, with the other he was defending himself against my blows.

Ludwik had already been to the station and found that no train was leaving for Stalingrad that day. He had also seen Viera. His suspicions had proved correct. The commander of the *voyenkomat* had actually spoken to the chief of the hospital and my disappearance was discovered in the afternoon. The search had begun immediately. Viera's superiors blamed her for letting me go and for furnishing me the uniform. She explained that she saw nothing against the rules in this, and that she had known I wanted to go to the *voyenkomat* to get some information about my transfer to the Polish Army. She had not suspected that I could be a deserter. Fortunately, Bierdieyev's papers were not mentioned. Ludwik assured Viera that I was in a safe place. She sent me her best wishes for success and told Ludwik she could not risk seeing me, as she feared that her every step was now being watched. Hearing that she was safe, I breathed a sigh of relief. At least I did not have to worry about her.

Ludwik and I then began to analyze the situation. It was clear that they would continue the search for a long time and that we'd have to count on more than one visit to this house. The town was small, and there were many patrols. Ludwik maintained that they would soon be sending out "Wanted" posters with a physical description of the deserter and a reward for information that could lead to my arrest. Only now did I remember that while in hospital I'd had a picture taken for my identification card, figuring it might come in handy when falsifying papers for myself. These pictures I had never received, as they were not ready. Now they would serve as excellent material for "Wanted" circulars. There were great numbers of such posters about deserters, with and without photographs. I had seen them in the waiting room of the *voyenkomat* and in the post office.

Ludwik thought that the train stoppage should work in my favor, that by the time they started again the excitement over my desertion would have calmed down. When they did start running, said Ludwik, he would arrange for me to be put on a train here in N—, where the entire garrison knew him. The military commander of the station, a lieutenant of the NKVD, owed Ludwik a few hundred rubles, and his eyes could be closed at the right moment. Ludwik, in fact, could and did help a lot. Although the landlady and her cousin took turns watching for patrols by day, I always felt safer at night when he was in the house. With his prestige he could prevent the patrol from searching, taking the responsibility on himself that nobody was hiding in the house. Nevertheless, I had my qualms.

"Ludwik," I asked him one day, "what would happen if we were both caught?"

Like myself, Ludwik was a fatalist.

"In that case," said he, "we'd both go to the wall."

Though the house was searched again on four occasions, I always managed to get to the cellar in time. Once there, I felt pretty safe. My hiding place between the cabbage barrels was excellent. Ludwik spent much of his free time watching over me and storing supplies of food for my trip. It was difficult to figure out how long this journey was going to take. On the map, the ground I had to cover looked impressive. A rough guess made the distance about two thousand kilometers. On crutches and on the run, that was a long way.

One morning (it was February 7, a date I'm not likely to forget) Ludwik went off by sled to inspect food storage houses somewhere in the neighborhood. The landlady also left. I was in bed, reading some old newspapers and from time to time looking out of the window in case a patrol should appear. Unfortunately, some news I found in a paper proved a little too interesting, for next time I looked out of the window, a patrol composed of four noncoms and an officer was at the gate. The situation seemed hopeless. There was no time to drag myself to the cellar. The landlady's cousin, terrified, burst into my room.

"Patrol!" she shrieked in a strangled voice. If they were caught hiding me they would be shot too.

Instantly I acted on what I prayed was an inspiration. In a hushed voice I told her to hide my crutches under the bed and inform the patrol that "only the major is at home and he's asleep." Meanwhile I quickly put on Ludwik's coat with the major's rectangles on the collar, spread a blanket over my legs, placed myself on my stomach and the insignia in a prominent spot in the pillow.

★ 201

I listened to the knocks on the front door and the old lady tripping across the floor to open it. The sounds of the ensuing conversation were dimmed by the closed door. Then the tramp, tramp of heavy boots mounted on the staircase. A knock on my door. I didn't answer. Another knock—louder. With a sleepy voice I let out as calmly as I could:

"Who's that? Come in!" and added a few curse words.

The door opened and I saw the face of a young lieutenant. His eyes glued to "my" insignia, he sprang to attention, saluted and recited:

"I beg your pardon, *tovarisch* major. I am under order to search houses for deserters. The landlady told me that you were asleep, but I had to make sure."

He then carefully let himself out into the corridor, closing the door silently behind him.

"To hell with all your deserters!" I shouted after him. "What's it got to do with me? You woke me up!"

I listened to them go through the entire house, down to the cellar, and then out of the gate. The old lady came into my room, trembling with fear. She still wasn't sure the danger was over. Nor was I, but I didn't tell her. Instead, I drank a tumbler of vodka.

When Ludwik returned in the evening he showed great nervousness at what might have happened. Then he told me that the trains were to start running within two days at the latest.

★ ★ ★

LUDWIK'S information was correct. On the morning of February 9, after I'd spent a week in hiding, he returned from the station with the news that at 1:30 sharp a train would pass through N— in the direction of Stalingrad. I was to go to the station on a sled and remain on it till the train came in. Meanwhile Ludwik was to inform the NKVD lieutenant who commanded the station that one of his own men, bound for hospital in Stalingrad with a broken leg, had to be put on that train. The undertaking was highly risky, but Ludwik was sure of himself, counting on the respect he enjoyed in the garrison.

When the sled arrived, four of Ludwik's trustworthy men wrapped me up in a fur coat and packed me onto it. Ludwik got in beside me and away we went, I holding onto my *nagan* in the pocket of my sheepskin jacket. On arriving at the station, a group of soldiers, noticing Ludwik, sprang to attention and saluted. I remained on the sled, watching Ludwik talking to the lieutenant of the NKVD.

Presently I heard the sound of the train. I stuck out my head. Even at a distance it was a gruesome sight, surpassing all my fears. It was so densely covered with human beings—on the roof, the steps, the bumper—that the train itself was invisible. With the thermometer at twenty degrees below zero, these people were arriving at their destinations with frozen noses, hands, and feet. Quite often they were frozen to death. I was filled with terror. What hope had I to get on such a train. I looked at Ludwik. Seeing my terrified eyes, he signaled me to remain calm. As the train halted I watched the lieutenant giving the train commander instructions. The soldiers of the station guard clustered around Ludwik. I could not understand what was going on. Suddenly I saw Ludwik stride toward a so-called "soft" coach [Russian trains have two classes: (1) "soft," meaning coaches with upholstered seats; (2) "hard"—coaches with wooden seats]. Standing in front of it, he slowly drew his *nagan* from its holster, waved it carelessly in the air, and then, aiming it at the coach, shouted in a stentorian voice:

"Everybody out of this car. Off with the baggage! Fast!"

The interior of the coach resounded with a roar of curses. The station soldiers rushed up to its four doors and without a word of explanation began pulling passengers off the train onto the platform. Meanwhile the NKVD lieutenant, standing next to Ludwik with his hands on his hips, rocked with laughter. It was hard to believe that so many people could have packed into one coach. Within a few minutes at least one hundred and fifty men and women were standing, glowering, on the platform. Soldiers then boarded the train to make sure no one was hiding under the seats. Suddenly, in a voice of thunder, Ludwik yelled a command in the direction of the station, whereupon two elderly women appeared with pails and brooms and climbed onto the train. After a while clouds of dust, dirt, and all kinds of refuse came pouring out of the doors. Then another woman appeared with a pail of carbol and began sprinkling the disinfectant throughout the coach. When at last the coach had been properly cleaned, Ludwik turned and surveyed the crowd on the platform. He let his eyes rove from face to face. Finally they came to rest on a pleasant-looking lieutenant, whose left arm hung in a sling.

"Where are you going, lieutenant?" I heard him ask.

The lieutenant was going to Kuibyshev.

"Would you like to take care of my sergeant who can hardly walk?" asked Ludwik.

When the lieutenant nodded, Ludwik told him to follow him onto the train. A few minutes later he reappeared in the door and called to his men on the sled. They promptly grabbed me as I was, wrapped up in my furs, and carried me like a package to a seat in the compartment pointed out by Ludwik. Then they left. On the opposite seat sat the lieutenant, staring at me with wide-open eyes. Just as he was about to say something, we heard Ludwik's voice outside, and immediately the thunder of feet as the crowd began pouring back into the coach. As they fought for seats, the coach literally shuddered under the onslaught. Although within five minutes all of them had managed to get into or onto the coach, nobody entered our compartment. With the help of the lieutenant I got up and stood at the window. The only people left on the platform were Ludwik, the NKVD lieutenant and his men, and the women with their pails and brooms. Ludwik's trusties were still standing by the sled. The engine gave a hoarse whistle; Ludwik lifted his head and saw me in the window. He smiled at me joyfully and saluted. I answered in the same way. The engine jerked, moved forward, and in a minute Ludwik had disappeared.

I lay down on the seat again. The lieutenant and I began to talk. He was an infantryman called Yudkov. His wounded arm had been so badly set that he could not straighten it out from the elbow. He was on his way to Kuibyshev where he had a job in an ammunition plant. I told him that I was going to Kuibyshev too, possibly further. I also explained that Ludwik had been the commander of my unit at the front and that this was why he had taken such affectionate care of me. Yudkov asked no questions.

The train went all the way to Stalingrad without stopping. The station was completely blacked out; only here and there little blue lamp bulbs showed the platform exits. Yudkov took my bag and we dragged ourselves to the hall to find out about a connection to the north. The information booth was besieged. The entire station was packed with men waiting for trains. They lay around on the foul, refuse-strewn floor. There was no free spot to be found. There were families evacuated from the German-occupied territories, men in uniform returning from hospital, men headed for some military courses or on furloughs. We were standing at the information booth, waiting our turn in the stifling atmosphere, when a general commotion of yelling and cursing broke out. A few men with Red Cross bands on their arms ran toward the disturbance. A few minutes later they returned, carrying two bodies on a blanket.

"Typhus," muttered a pilot standing beside us.

Typhus was reaping a rich crop at this time. It was particularly prevalent on the railroads, where people—sometimes traveling for weeks without washing, covered with lice, not realizing they had caught the disease—died within twenty-four hours. Trains were supposed to take on only passengers possessing a certificate proving they had gone through the "delousing" process. This was the only protection, but unfortunately insufficient, since those travelers who got on at smaller stations did not go through this procedure.

Since there was no train in the direction of Penza until the following afternoon, Yudkov and I went to a military food store and took out rations for three days. They didn't amount to much: a handful of biscuits, hard as stone, a piece of sugar, a few herring, two cans of kasha, and some stale tea. Since we did not feel like spending the night among lice and typhus germs, we decided to try to find a place to sleep elsewhere. It was well past midnight when we left the station. Stalingrad lay silent in the darkness. Crossing a bridge over the tracks, we passed some small houses in the station district. After a while we saw a gleam of light behind a partly closed shutter. Yudkov knocked at the door. The window was opened slightly and the frightened face of an old woman appeared.

"Mother," cried Yudkov merrily, "two soldiers want to spend the night here. Got any room for us?"

After a short discussion, the door opened. Our hosts, an elderly couple, poorly but cleanly dressed, beckoned us in. We put our bags in a corner and took off our sheepskin coats. The old people asked no questions. They ran here and there, bringing two basins and towels, then a pitcher of hot water. The old woman held me up while I washed. After a steaming samovar had been placed on the table, they began to talk. For forty years the old man had worked as a photographer in a large sanatorium outside Stalingrad. Now, however, the sanatorium had been converted into a military hospital, and since the beginning of the war photographic equipment had been unobtainable. So now the old man did nothing.

While pouring tea his wife declared in an embarrassed voice that unfortunately they had nothing to offer us but potatoes. We thanked her.

From Ludwik's supplies I produced a loaf of bread and a can of condensed milk, and offered them to the old people. They ate with great respect, admiring the sweet milk, with which they smeared their bread.

"We have not eaten bread for nine days," said the old man.

We felt uncomfortable.

"But why? Don't you have your rations?" asked Yudkov.

He sadly shook his head.

"He," said the woman, pointing at her husband, "is not a working man any more. When his job in the sanatorium ended, they never gave him the old-age pension which had been due him for a long time."

"But that pension should have come to you automatically," said Yudkov.

The old man nodded in embarrassment.

"Yes, yes, but you see, every time I go to the office there are always so many people, so many servicemen's families, that I can never reach my turn on the line."

We looked at each other.

"Well, well," sighed the photographer, "we have a few potatoes from our own little garden. Some people have nothing at all."

When we had finished our meal, the old couple prepared a place for us on the floor, close to the kitchen wall. Lying under our sheepskin coats, we were soon fast asleep.

It was still dark when Yudkov woke me to say that in case there should be an earlier train, we ought to leave immediately for the station to get our "delousing" certificates. The photographer had already gone to make one more effort to get his bread ration. We said good-by to the old woman, leaving her the bread we had broken the night before.

The station bath building was literally besieged. Yudkov now displayed his shrewdness. Grabbing me by the waist, he lifted me up and started pushing through the crowd, calling:

"Make room for a wounded soldier!"

Without a word of protest, the crowd made way and let us pass through all the way to the gate. In a large hall, where people were undressing and standing naked in line for their showers, we ran into trouble. If we gave our sheepskin coats to the *deskamera* they would be burned, or at best shrunken so as to be unwearable. This was a serious problem, but again Yudkov took the initiative. He solved the problem by surrendering his uniform to the bathhouse attendants and taking his shower while I guarded our sheepskins in the waiting room. When he returned he guarded the coats while I, keeping my equilibrium in a circus manner, stood under the shower on crutches, washed, and then joined him. Having waited, stiff from cold, a whole hour for our uniforms and underwear, we dressed and got on the last line to receive our certificates. There a fellow with an insolent face on a fat neck pointed a pudgy finger at our sheepskins.

"Have they been deloused?" he asked.

"Yes," muttered Yudkov.

The man gazed at him suspiciously and fingered one of the coats which, of course, was cold.

"That's a lie!" he barked. "They have not been through the *deskamera*. Leave them here and pick them up tomorrow."

But Yudhov was no fool. He knew this meant we'd never see our coats again. So he calmly drew his *nagan* from its holster and banged its butt on the table at which the employee was sitting.

"The coats are disinfected!" he hissed.

The employee paled. In silence he filled out both certificates and, without raising his eyes, handed them to Yudkov.

"Happy New Year!" mocked Yudkov.

He put the *nagan* back in its holster, and I followed him onto the street. Behind us an elderly attendant came out, his face wrinkled in a smile.

"You gave him a fright that time, *tovarisch* lieutenant!" he laughed. "That's how he steals coats every day. Not everybody answers him the way you did."

While making our way toward the station, Yudkov rolled a couple of cigarettes, put one in my mouth, and handed me a light.

"Are you going to the Polish Army, sergeant?" he casually asked.

I almost choked on the smoke. How in hell could he know that? He looked at me calmly.

"While waiting for you to come out of the shower," he said, "I quite automatically looked over your papers."

I could have beaten myself for my stupidity. How could I have left my papers in my coat? I tried to speak very calmly.

"Yes, I am going to Buzuluk. I'm Polish "

"And those other papers, whose are they?"

My mind was working feverishly.

"Mine," said I, suddenly struck by a brain wave. "In the army I was known by a pseudonym. My real name is on the document directing me to Buzuluk."

The explanation was ridiculous, since I had a whole set of documents issued under the name of Bierdieyev.

Yudkov was smiling.

"You don't have to explain to me," he said, shrugging. "It's not my business."

We walked on in silence. I was completely thrown off my balance. Purely from my own carelessness, I could now be caught at the very beginning of my trip. In the station Yudkov looked me in the eyes.

"I tell you again," he said, "don't worry that I happened to discover the object of your journey and the double documents. It's none of my business."

His tone of voice suggested that one could count on his discretion.

In the station I sat down against a wall and watched over our bundles while Yudkov went to see whether by any chance there was a train leaving for Penza. I was sitting there, observing the station traffic, when I began to realize fully the colossal risk connected with my plans. If Yudkov were a zealous man, there was no doubt he could make use of what he had seen—and my trip would end right here. I had always known what threatened me if I were caught, but only now did I properly appreciate the meaning of—"a bullet in the head, without trial." Any patrol smelling a deserter in me had the right to shoot me at the first fence. I gave the butt of my *nagan* in my pocket a good squeeze, deciding never to let it out of my hand. Should I fall into the hands of the NKVD, I would always have time to use it. I would not let them take me alive.

Suddenly I saw Yudkov dashing toward me.

"Our train has been here all morning," he said, panting for breath. "It's already full. I'm afraid you'll have to try to cross the tracks. We must get into it at once."

I limped after Yudkov as fast as I could. Actually, the train was on a side track outside the station, and not due to appear at its scheduled platform until half an hour before its departure. The crowd now on it had discovered its existence by the custom, prevalent both in war and peace, of bribing the railroad employees. By the time I managed to reach the train, Yudkov had already found space for us both. With enough room to half sit, half lie, I fell asleep immediately.

I spent the five-day journey to Penza partly asleep and partly in conversation with Yudkov. The coach was terribly crowded. The closer we came to Penza, the more crowded it grew. Fortunately, our invalid privileges were honored and we were fairly comfortable. We slept next to each other, our bags under our heads. I grew so used to squeezing the *nagan* in my hand that I almost forgot I held it. But my legs felt worse. They had begun to hurt me already in N—, but now I felt that under the plaster they were full of pus and that the wounds must have opened.

The train dragged unbearably and often made long stops in the open fields or at small stations. At the bigger towns where there were Red Cross dressing stations, it stopped but a few minutes because of the crowds of waiting passengers, which even the military guards could not hold back for long. Finally at one of the long stops which the train made at a little station, Yudkov searched the train for a doctor. He at last found an elderly colonel who tapped the plaster cast, smelled it, and decided that since he had no disinfectants or bandages with him, he could be of no help to me. He advised me to leave the plaster casts on in spite of the pain, since they would at least keep my legs immobilized during the trip, which was the important thing.

On arriving at Penza, we discovered that our train continued north as far as Kazan. Yudkov went to inquire for a connection going east to Kuibyshev, while I sat against a wall and watched the station traffic. I think there was more of it and that it was smellier than that at Stalingrad. I certainly saw more victims of typhus. I counted six corpses on blankets by the time Yudkov reappeared. The train to Kuibyshev, he said, would not arrive for two days. The infirmary where I wanted to have my legs seen to was so overcrowded with people dying of typhus that Yudkov advised me strongly not to go near the place. He also thought it advisable to quit the station to avoid lice. I do not mean to suggest that the two of us were louseless. On the contrary, we were dirty, unwashed since Stalingrad, and so covered with lice that when we reached a hand under our shirts to punish one irritating louse, we brought out a handful of at least twenty— big brutes, drunk with blood—which we threw away, since it was impossible to kill them all. But fortunately our lice were the good old healthy companions of the soldier's life, none of whom had previously bitten a typhus patient.

★ ★ ★

WE left the station and went in search of a place to sleep. It was not an easy job, as the city was packed with refugees. We knocked at many doors, but wherever we went we met with a cold and indifferent refusal. We were about to abandon the search when Yudkov gave what we decided would be a last knock at the door of a miserable little house. The door opened wide and a woman of about thirty appeared on the threshold. In a thick, expressionless voice she asked what we wanted. Yudkov repeated his eternal petition for a place to sleep. She nodded, and with a gesture invited us in. On the dark porch I was struck by a smell of mold and dust. We entered the kitchen. On the table stood a broken bottle with a candle-end smoking from its spout. The shutters were closed and the room was very cold. Everything was covered with thick dust. I wanted to sit on a stool by the table, but the woman stopped me and wiped the dust off the stool with her apron. She raised her head and smiled sadly.

"I've not done any cleaning for a long time," she said.

Yudkov and I looked at each other in surprise.

"Six years," she added, as if to herself.

Yudkov started to ask her something, but she interrupted him.

"Have you come a long way?"

"From Rostov," answered Yudkov laconically.

"From hospital? Wounded?" inquired the woman.

I nodded.

"Would you eat something warm?" she asked in a suddenly friendly voice.

"We don't want to trouble you," began Yudkov.

She waved her hand.

"No trouble. Only you will not get anything special. I will make soup!"

She began looking around the kitchen, searching for something. After a while she let out a curse, grabbed a stool, and, placing it against one knee, smashed it into several pieces. Then she took an ax and cut one of its legs into small bits for the stove. Yudkov handed her some matches. She mumbled something that might have meant thanks, and lit the fire.

Avoiding our eyes, the woman then found a pot in a kitchen closet, rinsed it with water, and began to peel potatoes from a small sack by the stove. Yudkov got up.

"Shall I bring you some more wood?" he asked in a friendly voice. "Otherwise you will burn up all the furniture before the soup boils!" She laughed soundlessly, displaying her teeth.

"It would help," said she, "but I don't think you will be able to find any around here."

Yudkov answered her with a smile.

"Not far from here I saw a wooden fence. Fences burn well." He put on his coat and went out. She dropped the potatoes into the pot, produced from under the table a small bag of flour, and made dough for some noodles. Without turning her head from the stove, she asked:

"What's wrong with your legs?"

"Wounded," I answered. "And not healing too well."

She was nodding. .

"Well, will we win this war or lose it?" she asked.

I was startled.

"We will win, of course." I said without conviction.

She turned her face to me.

"You can speak freely. Do you believe what you just said?"

I shrugged.

"It is difficult to say for sure, but one has to hope."

Her face again acquired a severe expression.

"Nobody has ever conquered our sacred Mother Russia. The Germans won't be able to either."

I didn't know what to say.

"Well, it's all right," she laughed. "I'll give you a certificate stating that you are a 'reliable' citizen "

I didn't have the time to open my mouth when Yudkov returned with an armful of boards torn out of the fence. He began cutting them up and throwing them onto the fire. The woman was definitely becoming gayer.

"Weren't you afraid that the militia would catch you?" she asked.

He laughed.

"Fortunately today a militiaman would be afraid to stop me. It was different in peacetime."

The room was warming up. From my bag I produced a can of meat and threw its contents into the pot. Yudkov finished cutting the wood and turned to the woman:

"It looks as if you haven't been home for a long time."

She looked at him.

"Didn't I tell you that I haven't been home for six years—that I came back this morning?"

"This morning?" repeated Yudkov in amazement. "Well, no wonder you have no household!"

He looked at the woman with interest.

"And where were you all this time, if I may ask?"

She hesitated. Finally she said, "In prison."

"For what?" he asked.

"I cannot answer you," she said, "because I don't know."

She remained silent for a while, rolled a cigarette with the tobacco Yudkov offered her, and told us her story.

She was from the Ukraine. Eight years ago she had graduated from the Teachers' Seminary in Kiev and was given a teacher's position in a *diesietiletka* (ten-year school) here in Penza. At the end of the second year of teaching she had some difficulties with a few pupils in her class. One night men from the NKVD suddenly came and arrested her and put her in prison. She was there for three months without even hearing what she was accused of, then there was a short "trial." She heard with terror that she had poisoned the minds of the children in school by anti-revolutionary propaganda. There was no explanation possible. When she swore that she knew nothing of this and that it was surely a mistake and slander, they threw her into solitary confinement where she was "interviewed" for several days and nights. During that time they knocked out several of her teeth and broke two ribs. Finally, unable to stand any more, she had "confessed" sins she had never committed. The verdict: ten years jail for "anti-revolutionary activities." She'd been thrown into prison at Wyazma. Thanks to an iron constitution, she had managed to withstand its frightful conditions. When the Germans approached Wyazma, a part of the prison was evacuated to the east. The rest, of whom she was one, were freed for "correct conduct." She had spent six years in prison. When liberated, she was told that her arrest had been caused by two of her pupils denouncing her. Since then she had been wandering around, working on the fortification of defense lines. She had returned home today. During these six long years her home had been locked and stamped by the NKVD. Usually a house stamped by the NKVD was eventually given to some "inside" henchman. But the teacher's home was so poor and old that nobody wanted it. She had entered it two hours before our arrival, bringing into it her broken life and her food supplies—a small bag of flour and a small bag of potatoes.

For Yudkov this story was nothing new. As for me, even though I had often heard such things, it was the first time I had met such a case directly, and I was very shocked. I was that much more surprised by her faith in victory.

"It is admirable that in spite of these sufferings your patriotism should remain untouched," I remarked.

She raised her eyes to me.

"It depends," she said, "on what one understands by patriotism. I personally mean love for the fatherland, for Russia, which has nothing to do with the people who were guilty of my misfortune or with the people who are sitting on top."

"Enough of this," said Yudkov. "Let's have some of that soup."

We ate the soup in silence and began to prepare ourselves for the night. Yudkov brought in some more wood and made a big fire in the stove so that we were sure of keeping warm till morning. I wrapped myself up in my coat, put the *nagan* under my head, and fell asleep.

When I woke I could see sunlight through a crack in the shutters. I got up and opened a window. By the stove, covered with blankets, Yudkov and the woman were lying side by side. She opened her eyes, gazed around sleepily, saw me, smiled, yawned, looked at the snoring Yudkov, and promptly fell asleep again.

I went out onto the street. I wanted to find a bazaar and buy something to eat. I had two hundred rubles which Ludwik had given me before my departure. During the past few months I had not been able to spend any money, since there had been nothing to buy. Here, however, farther from the front, I hoped to be able to put these wretched pieces of paper to some use. When I finally found a bazaar I discovered that everything could be bought, but not for rubles, only for barter. For smoked ham, at the sight of which my mouth began to water, they wanted my boots. For a large round loaf of bread, they expected my trousers or blouse. Thus prevented from spending a single ruble, I limped toward the station to find out something more specific about our train. On the snow-covered square in front of the station I noticed some large boards adorned with multicolored posters, announcements, orders, information bulletins of the agency Tass.

Approaching closer, I leaned against a lamppost and rolled a cigarette. "Death to the invader!" screamed the posters. "Everything for the front!" Then a man appeared with a roll of fresh posters in one hand and a pail of paste in the other. He put the pail down and began slowly tearing off the posters that were out of date. Tired, I remained leaning against the post, puffing the stinking cigarette and watching him work. He dipped the brush in the paste and smeared the prepared board. Then he picked out the bulletins from his roll and began placing them on the board one after the other. They had nothing new to say: "No changes at the front; our armies occupied a few villages, capturing so and so many guns, tanks, and planes. Our losses—negligible." Having finished pasting the bulletins, the man now produced from his roll some smaller posters which he started pasting side by side. Though the print was small, I immediately read one large word which electrified me: "DESERTER!"

I drew closer and began to read. These were "Wanted" posters for the arrest of deserters, some supplied with photographs, others only with descriptions. They all contained the same information, adding that aid in capturing a deserter carried a five-hundred ruble reward. The man had pasted up about fifteen posters, when he unrolled one which almost made me shriek out loud. My own cynically smiling eyes were staring at me from the paper! Another picture showed me my own profile, still unfamiliar to me because it had been taken after I broke my nose at the Dnieper front. Suddenly I felt dizzy. I saw darkness in front of my eyes. I leaned more heavily against the post. The cigarette was burning my lip; I spat it out and tried to read the text of the poster, which was jerking up and down before my eyes. Above the pictures, in large letters, I saw "500 RUBLES," then a mimeographed form about the desertion from the hospital in N—, a description of my possible uniform, which also mentioned the sheepskin coat, my name (for once spelled correctly), my age, appearance, and details about my wounded legs and the crutches. At the bottom, as the authority responsible for issuing the description, I read the signature of the hospital commander in N—.

Even though I had counted on the possibility that the pictures taken at the hospital would be used for identification circulars, the actual sight of them was overwhelming. I read the description of myself three times; it was pretty thorough. I began wondering how this poster could have caught up with me, since we had caught the first train out of N— after my flight and I had not missed a train in Stalingrad. Could it have traveled with me, on the same train?

I hobbled into the station. The sight of the first NKVD patrol made my heart race. I was convinced that everybody had read the posters and that they were on the lookout for me alone. But the first patrol passed me by indifferently. So did the second and third. Then a sergeant from the fourth came up to me and, in a friendly way, asked if I needed his help. My throat felt parched as I inquired about a train to Kuibyshev. The sergeant informed me that the next transport to Kuibyshev would leave the following afternoon, and he added that this morning a train from Stalingrad to Kuibyshev had passed through, but that it had been a "special" and so packed that it would have been out of the question to board it. Well, that must have been the train on which the posters had come. What was worse, they would no doubt precede me wherever I went! I thanked the sergeant and was about to leave when he casually asked me for my documents. Instantly, I was completely calm. The nervous tension had passed away. Nevertheless, I gripped the *nagan* in my pocket with my right hand, while with my left I handed the sergeant Bierdieyev's papers. He glanced at them and handed them back.

"It's my duty, you see, to check the documents of single soldiers," he explained. "There are masses of deserters about. But if anyone at H.Q. thinks it's easy to catch one, he's mistaken."

I smiled kindly.

"This must be a very responsible position, sergeant," said I.

He waved his hand, saluted, and went away.

Since there was now no reason for me to stay in town, I set off toward the house where we had slept. On the way I kept thinking of my situation. In principle nothing had changed. I knew from the beginning that the chance I was taking was tremendous; but the shock I had just received had not really shortened the odds against me. The average man is not particularly interested in desertion posters. I had no need to fear those not directly concerned with the catching of deserters. As for the patrols, the NKVD, the railroad station crews, they did not seem so dangerous. I reminded myself of the reward offered. Once in Odessa I had received seven hundred rubles for my watch. I began laughing like a lunatic. What would my friends say if they knew the price on my head!

In front of the ex-convict's dirty hut I stopped. What of Yudkov? What would he have to say when he saw my face on the poster? It would be bound to catch his eye. As a "reliable" citizen, it would be natural for him to denounce me. What could I do about that? Nothing—nothing but rely on luck, which so far had been good to me. I squeezed the handle of the *nagan*, which was my last resort if luck turned against me.

The idyll between Yudkov and our hostess was still going on in the hut. Neither seemed particularly disturbed by my entrance.

"Any news of a train, sergeant?" asked Yudkov.

"It's not leaving till tomorrow," I told him.

Yudkov seemed as satisfied with this news as I was with the fact that he had not yet seen my face looking at him from a wall.

Next morning, after Yudkov had changed the date on our disinfection certificates to that day, we said good-by to his girl friend and set off to get our three-day food supply before catching the train. On the way we passed right under the board to which my face was pasted. I watched Yudkov carefully, but he didn't give it so much as a glance. Again in the station, in a most honored spot, beneath the portraits of Stalin and Lenin, hung a row of posters, among them mine. But Yudkov didn't raise an eyebrow.

We waited outside the gate to the platform where the train was due. Suddenly a human mass was pushing in our direction. Several NKVD lieutenants standing at the gate began to let people through singly. I saw that they were examining papers. Moving with the crowd, we were almost at the gate. Suddenly Yudkov seized my arm and called out:

"*Tovarisch* lieutenant. Here is a wounded sergeant on crutches." The officer looked at me over the sea of human heads and beckoned to us. We pushed our way through the crowd to the gate and faced the crossed rifles of the guard.

"Papers," said the lieutenant.

He looked up as he handed mine back to me and said: "Haven't we met before, sergeant? Your face is very familiar." I was certain that I had never seen him before and that my face was probably familiar to him from the poster.

"I also remember you, *tovarisch* lieutenant," I said boldly. "We must have met somewhere. Have you ever been in Odessa?"

"No"—he shook his head—"I have never been there. It must have been somewhere at the front."

And then we were passing through to the train.

For once the train was not overloaded. There were even several empty berths in our coach, which was partly occupied by people belonging to the privileged class. Among them were many women and children, all quite well dressed. Instead of the usual bags and baskets, the luggage racks were filled with suitcases and trunks. The entire coach, which was not divided into compartments, gave the impression of one large family whose members were taking a long trip together simply for the fun of it. Our entrance, therefore, caused quite a sensation. I on crutches, Yudkov with his arm in a sling, both in our sheepskins, we no doubt looked a little out of place. We stopped, undecided, in the doorway, not knowing where to go. Then an elderly man, a major with a doctor's insignia, came toward us.

"Good day!" said he, saluting. "Where are you going to, comrades?"

When we had introduced ourselves, the doctor helped me on to an upper bunk next to a woman with two children. Yudkov was given a berth on the same level; I could see him through the wooden rods which separated us. I took off my coat without letting the *nagan* out of my hand and put it in my trousers' pocket.

Presently the doctor, established further along the coach and below us, called us down for a glass of vodka. This was an evacuation train from Moscow, he told us—a rare thing nowadays, for since the December counteroffensive such trains had almost ceased to run. Our fellow passengers were families of higher government officials, of staff officers, or of officers going either to the far east or south for military courses. In a sense, they were the cream of Moscow society. Their manners and speech were certainly very different from what I had been used to. Beside us on the floor a couple of prettily dressed children were playing with a toy train—something I had never seen in Russia. As the engine ran against my feet, I picked it up to wind it. Suddenly it struck me as oddly familiar. Then I saw, stamped on its front: "Gerngross, Wien." As a child, I'd had one just like it.

"My husband," said the children's mother, catching my eye, "brought that back from Lwów. He was stationed there several months."

Yudkov promptly let out that I was Polish. Seeing my worried look, he waved his hand reassuringly and added in a loud voice: "That's all right, sergeant. After all, you're traveling legally and you have an official transfer."

This news created a great sensation. I was immediately showered with questions. After I had given them a brief account of my war history, the major told me that he had seen General Sikorski's Polish mission when it had come to Moscow on Stalin's invitation a few months ago. And he added that though he was not positive, he was under the impression that the Polish Army which had spent the winter near Buzuluk and Totskoye, was now in the process of moving to Tashkent and Samarkand, in southern Siberia. This news surprised and worried me. If true, it meant that my journey, instead of ending in Buzuluk, would be prolonged at least another thousand kilometers. The major, however, consoled me with the information that our train was going all the way to Tashkent and that I wouldn't even have to change.

With this comforting thought, I began to prepare for the night. Yudkov had helped me onto the upper berth and laid my crutches by my side, when he smiled suddenly and let out in a whisper:

"I thought you came out very well on those pictures at Penza station!"

I gazed at him, speechless, I watched the smile fade from his face.

"Don't worry," he said. "I haven't seen anything. After all, since looking over your papers in Stalingrad, I expected this. As you see, I haven't said a word."

Then he climbed onto his berth, waved a good night, smiled, and a minute later was snoring.

★ ★ ★

WE reached Kuibyshev on February 20, after two days in the train. This was the end of Yudkov's journey. As we shook hands for the last time, he leaned toward me.

"Your army," he said, "is supposed to be leaving the Soviet for England or the Middle East. When you catch up with it you can pay me back for not having given you away."

I was most surprised by this information, but even more by the suggestion that I could pay him back when I'd caught up with the Polish Army,

"I really don't understand you," I told him.

"Just tell people what you saw here," whispered Yudkov as he turned and walked away.

An hour later the train moved on, stopping at Buzuluk the following morning. The first thing I saw at the station was a group of soldiers in British uniforms and red-and-white armbands.

"Look!" shouted the doctor from the next window, "your people are still here!"

I grabbed my bag, shouted good-by to everybody, clambered down to the platform, and limped up to the first soldier in the group. He asked me in Russian what I wanted. With my eye on his Polish armband I actually forgot for a moment that I had on a Russian uniform, so I asked him in Polish if he could direct me to the Polish Headquarters. He looked at me sharply.

"So you're Polish," he answered in Czech. "You see that flag? That's where your headquarters are!"

At first all this seemed most mysterious. Then I noticed that the colors on his armband were inverted: the white underneath, and the red on top. The British-organized Polish units had already left, their places having been taken by Czech units, also organized by the British.

The flag to which the soldier had pointed, however, was definitely Polish (the red below the white). It flew from the roof of a hut behind the station. I pushed the door and went inside. I found myself in a large room heated by an iron stove. Along the walls on benches sat about a dozen men, and at a table near the window a captain and a lieutenant. They were all wearing prewar Polish uniforms. My entrance seemed to cause quite a flutter of surprise. I hobbled toward the table.

"Captain," I said, "I want to report for service."

Both officers gaped at me. Rising from the table, the lieutenant walked all around me, staring at me as if he were inspecting a horse.

"How do you know Polish?" asked the captain.

"My mother taught me," said I, smiling.

"Ah," concluded the captain, "Polish mother, Russian father!"

This was too much.

"I am from Cracow," said I, "a student of the Yagellonian University." And from the depths of a pocket I produced the only Polish document still in my possession—my university identification card.

The two officers stared at it in disbelief. The captain was also from Cracow. He even knew my father! After that everything went smoothly. I briefly explained my presence and the reason for my uniform. Yes, they had seen large groups of boys from the Red Army pouring in, but never a sergeant in a complete field outfit. The officers then informed me that the Buzuluk units had left here, and advised me to go on by train to a place called Vrevskoye, beyond Tashkent, where regular inductions would be made on February 28. Glancing out of the window, I saw that my train was still in the station. It was bound for Tashkent. Without a second thought, I shouted good-by, hobbled out of the hut and into the street, horrified that I mightn't make it. But I did, by a split second. The doctor and my other acquaintances had seen me through the window. When I gave them my reasons for continuing the journey, the doctor nodded.

"Ah, that's better," said he. "Down in the south there's no winter. Soon everything will be green!"

I spread myself out on a seat and began examining the map of Russia which the old photographer in Stalingrad had given me. I figured that Tashkent was about one thousand kilometers away. Borrowing a pen from the doctor, I calmly wrote Vrevskoye as my destination on all my documents. Then I supplied each of them with an illegible signature.

That night my legs hurt so unbearably that when we reached Chkalov the following morning I decided to go to a dressing station and ask to have the casts taken off. An official told me that the train was to make a stop of at least two hours, so I decided I had plenty of time. I was hobbling along the platform, looking for the infirmary, when out of the corner of my eye I caught sight of a fresh row of "Wanted" posters! And again, in a most prominent spot, I saw my cynically smiling eyes. Hurrying as fast as I could I turned in at a door marked with a red cross, passed down a corridor, and stopped at the entrance to the infirmary. I was about to knock when it suddenly occurred to me that maybe under these circumstances I'd better give up the idea of a dressing and get back on the train. At that moment, however, the door opened and a nurse beckoned me inside. I followed her in, sat down on a chair, and explained to her my reasons for coming. Without a word she took off my *valonki* and with great difficulty began cutting through the plaster. Finally it opened and fell off my legs. What I saw very nearly made me vomit. The wounds were open. The avitaminosis wounds were also active. My legs, covered with pus, were alive with hundreds of bloated lice. The stench was appalling. Turning away, I apologized to the nurse for exposing her to such a sight. She laughed.

"This is my profession," she said gaily. "Anyway, don't think you're the only one in this condition. Nearly all wounds are like this. Lice love warmth. A cast is one of their favorite hideouts."

She raised her head.

"Did you get these wounds at the front?" she asked.

I nodded. She lowered her eyes, but immediately looked up again and stared me straight in the face. Her lips tightened for a brief moment as though she were under tension, and I thought I noticed a look of revelation in her eyes. All this did not last longer than five seconds. Then she again lowered her head and finished cutting the plaster on the other leg. By her movements it occurred to me that she was working automatically, thinking of something else. In a flash I had it! All of a sudden I was positive she had recognized me from the poster. My muscles were so tense that I almost pulled the trigger of the *nagan* in my pocket. I was feverishly thinking what to do, when the nurse, clearly avoiding my eyes, got up from her stool.

"I've got to get some disinfectant," she muttered. "I'll be right back."

Still without looking at me, she walked slowly toward the door, as though preoccupied. My head was in a whirl; I could not collect my thoughts. I felt sure of one thing: the girl had gone to notify the NKVD. I glanced around the room. She had taken the only exit. My crutches were far away on the other side of the room. Despite the devilish pain in my legs, I walked a few steps and sat down on a stool in a corner, with my back to a wall and the door straight ahead. I was quite certain now that this was the end of my journey.

"Sergeant," said I to myself aloud, "keep your head to the very last moment!"

The sound of my own voice, strangely enough, instantly calmed me. I figured that about eight meters separated me from the door. They would not fire at me from there. When they came in I would do away with one or two and then shoot myself. Taking the *nagan* out of my pocket, I released the safety catch, blew a few breadcrumbs out of the barrel, turned the drum so that the revolver could be used immediately, held the weapon against my knee, covered it with my cap—and waited. But this waiting was terrible. It completely unnerved me. I tried to analyze the reason why. It was not fear; fear always produced a different reaction in me—the sensation of blood running from my body to my legs, giving me a feeling of numbness. This was something else, which I could only define as waiting for inevitable death. I had no doubt that I'd lost the game and that in a moment I would have to perish. And I think that this certainty, coupled with the waiting for the inevitable, caused my thoughts to chase each other. In a flashing sequence, images from the last three years of war began to fly before my eyes. They were broken up, disconnected, but very distinct. Five minutes had passed since the girl's departure. I thought of trying to escape, but soon abandoned the idea. Before I'd have the time to bandage my legs and put on the *valonki*, they would be here. And even if I managed to get out of the station, who could be easier to recognize and catch than I on my crutches?

I don't know how much longer that waiting lasted! It could have been ten minutes and it could have been half an hour. For me it was an eternity before I heard steps in the corridor. I couldn't be sure whether they were the steps of a single person or several. But the moment I heard them I promptly regained my calm and self-control. The door opened and the girl came in with a bottle of a yellow liquid in her hand. Seeing that I had changed my seat, she looked at me with surprise, and a swift glance of understanding passed over her face. She came up to me and, without asking why I had moved, began washing my wounds. I sat there, holding onto the *nagan* under the cap with the one hand, covering the cap with the other, and wondering what had happened. It crossed my mind that perhaps she had sent someone out to look for a patrol and had returned so as not to awaken my suspicions. Or else, maybe they were waiting for me outside, or ... I didn't know what to think. And again I began to grow nervous. Despite the pain inflicted by the disinfectant, I paid no attention to what the girl was doing to my legs. As for her, she worked on in silence without once looking up.

Only when she was through bandaging my legs, had put on my *valonki*, and risen from her stool, did she glance up and most unexpectedly smile.

"There's nothing to worry about, sergeant!" she assured me.

I said nothing. I just sat there, unable to make myself get up and go. At last I put the *nagan* back into my coat, the cap on my head, gripped the crutches, and limped slowly over to the door. She was standing in the middle of the room, looking at me. At the door I stopped. What if somebody were behind the door, waiting for me? How could I use the *nagan* under these circumstances? I again glanced at the girl, then at the door. She smiled slightly, dropped her eyes, and calmly said:

"There is no one behind the door. I told you there's nothing to worry about."

Leaning heavily on the right crutch, I lifted the left one and pushed the door. It opened with a loud squeak, revealing the empty corridor. Behind me I heard the girl burst out laughing.

"You believe me now?" she asked.

"One should never believe women!" said I, regaining my calm. I was still certain she had seen the posters and recognized me. Why didn't she give me away?

Not knowing what to say to her, I just mumbled, "Spasiba—thank you," and hobbled on down the corridor. At the end of it there was a door with a glass pane. Through it I could see the blue cap of an NKVD soldier standing with his back to me. Had the girl laid a trap for me after all? I stopped, not knowing what to do. Suddenly the soldier turned around and saw me. With a quick movement he opened the door and came toward me. For an instant I wanted to pull the *nagan* out of my pocket, but I saw in time that his face was not unfriendly. In fact, he was smiling gaily as he said:

"Wait, *tovarisch* sergeant, let me help you!"

With his arm around my waist, he led me carefully toward the platform. We hadn't covered more than a couple of meters when I thought I heard steps behind me, and turned around. The nurse was standing in the doorway, her face split by a broad smile.

"Bon voyage!" she called to me.

I thanked her. I also felt like smiling now.

"What train are you taking, sergeant?" asked the NKVD man.

I pointed to a long line of coaches at the end of the platform.

He helped me climb the steps, said good-by, and went away.

★ ★ ★

THE route south, through Kazakhstan, is wild desert country very sparsely populated. At the few stations we saw only Asiatics. Halfway between Chkalov and Aktyubinskaya we crossed the geographic frontier dividing Europe from Asia.

During this long train journey the passengers had time to become very familiar. Any show of modesty soon disappeared. Women washed and dressed without embarrassment in front of the men. When we managed to get a little extra water, everyone did his laundry in the coach; diapers, panties, petticoats, brassieres, were hung up to dry on strings across the aisle. By this I do not mean to suggest that we were clean. On the contrary, we were all devoured by lice, with no means to get rid of them.

Our more serious trouble, however, was food. In Chkalov, none of us had been given his food supply because the rations we had taken out at Kuibyshev on February 20 were supposed to last three days. It was now the twenty-third, but we were still without rations. Actually, no food was handed out in the normal way until we reached Tashkent four days later. To make matters worse, it was impossible to buy anything. Only barter. The female attendant on our coach had a store of flour, bacon, and peas in her compartment. For small amounts of these she would accept parts of our wardrobes, particularly winter clothing which, on her return to Moscow, she could sell for enormous sums. Since I had nothing left to offer, the doctor and the mother of the children took turns to keep me fed. Only once, at a small station beyond Kzyl-Orda, did we see anything for sale. A group of Kazakhs were offering some stale wheat cakes at astronomical prices. The doctor and I pooled our resources and bought one large cake for five hundred rubles—as much as the Soviet authorities were willing to pay for my head. In Soviet Russia human life is cheap, cheaper than bread.

We arrived at Tashkent just before noon on the twenty-seventh, after ten days in the train. I said a last farewell to my companions, all of whom were traveling further east.

Tashkent is a purely oriental city. Dirty, its streets covered with refuse, packed with thousands of refugees from the west, tortured by hunger, typhus, and dysentery, it made a very sad impression. I did not venture far from the station, as the train for Vrevskoye was to leave at five in the afternoon. In a military information booth, I found a Polish corporal in British battle dress, with a tin eagle on his cap. It was an hour's ride to Vrevskoye, he told me. Then he began warning me against thieves in the city. The military police and militia, he said, were helpless against this mass of robbers who were pouring in from all parts of Russia. They were stealing everything they set their eyes on.

"Well, they won't get anything from me!" said I, and set off to the mess, for which the Polish corporal had given me a ration card. Literally hundreds of men in uniform and civilian clothes were lined up outside the door through which came the smell of sauerkraut and potato soup. At this rate, I thought to myself, it will take me two hours to get something to eat. I had turned to go when a captain came up and asked me why I was leaving.

I explained that I hadn't the strength to stand two hours on my wounded legs and that I was afraid of missing my train to Vrevskoye.

"But you don't have to wait your turn," he said, leading me straight into the mess hall. "The wounded are privileged."

I put my crutches in a corner and seated myself on a bench. A young waitress with an intelligent face brought me a deep tin dish of some nondescript soup and a large slice of bread. Suddenly I realized I'd left my spoon in the train. The girl brought me a large wooden one.

"Keep it," she smiled. "You can't travel without a spoon!"

I thanked her, ate my soup, put the spoon in my pocket, and got up to go. By leaning against the table, I made my way to the corner where I had left my crutches. They had gone! I looked around helplessly. Not a sign of them. What on earth could anyone want with crutches? I didn't know what to do. Then the waitress approached me.

"What's wrong, *tovarisch* sergeant?" she asked cordially.

When I told her she let out such a string of curses that even I blushed. Then she turned and disappeared. In a few minutes she was back again with a stool.

"Sit on this," she said, and off she dashed again, through the crowded mess hall.

I had been sitting there half an hour, certain I'd never see her again, when suddenly she reappeared. In one hand she held a short crutch and in the other a cane.

"These are all I could find, sergeant," she said. "Try them; maybe they're better than nothing."

I tried them. She was just about right: they were agony, but better than nothing.

"Where did you find these treasures?" I asked her gratefully.

She shook her black curls.

"Find?" she repeated. "I stole them, of course!"

When I opened my eyes wide, she burst out laughing.

"Don't let that surprise you!" she said. "Yours were stolen too! You mustn't look for justice in Tashkent."

★ ★ ★

I REACHED Vrevskoye at dusk, to find crowds of Polish soldiers on the platform. When I asked the sergeant on duty for the draftees' assembly point, he was astonished to hear me speak Polish.

"Where are you from?" he asked.

"Cracow," said I.

The sergeant seemed very happy. He led me to an Uzbek wineshop, where we sat down at a table with a few other soldiers.

"There's a man from your town here," he said. "I'll go and get him."

He left the place and after a few minutes reappeared with a tall, thin, red-haired fellow. The moment the young man saw me he came straight up to the table.

"What!" he exclaimed, clapping me on the back. "You don't recognize me?"

It took me a full minute to realize that this was Tadek B., a childhood friend who had lived across the street from us in Cracow. We had graduated from secondary school together; then he had started his engineering studies in Lwów, so that I had not seen him since 1937.

When the sergeant and the others saw that we had a lot to talk about, they got up and left. Tadek told me that he had been given eight years in a Russian concentration camp when the NKVD had caught him trying to cross the frontier into Hungary in 1940. After the amnesty in 1941 he was freed, and since November of that year had been in an officers' school for tank units.

We then began discussing our native city—ancient, beautiful Cracow. Slowly sipping the sweet Uzbek wine, we let our imaginations wander together through the streets of the beloved city. Before long we were competing with each other in our memories, loudly capping reminiscence with reminiscence. For me this was no ordinary chat. As I talked I could feel the nervous tension flowing out of my body. With the wine coursing through my head, one thought dominated my mind: I was safe, free, among my own people!

Suddenly I felt I couldn't talk any more, so I just sat there listening. Tadek seemed to sense that something was happening to me.

"What is it, Fred?" he asked. "Are you not feeling well?"

I shook my head. I knew it was silly, but I could not control my tears.

"It's nothing, nothing, just a reaction," I assured Tadek. "Just the tension, you know, the constant fear of falling into the paws of the NKVD."

Tadek patted my back.

"Well, that's all over now. Forget about it. Here you're safe."

"I'm not so sure," I said, smiling. "Even here I'll have to be careful. My desertion can be found out. Don't forget—we're still in Russia."

He shook his red head.

"No, Fred, even if they find out they can't do anything to you. There have already been a number of cases like yours, and in every instance the Russian military authorities who have refused to issue the transfers in violation of the general order freeing Polish citizens from service in the Red Army have been severely reprimanded." I took my hand, holding my gun, out of my pocket. Tadek looked at it with interest.

"So now I won't have to hold onto this any longer," I said.

But I could not open my hand. I tried to pry my fingers loose with my left hand. I could not. Tadek tried. My fingers were cramped so tightly around the gun that later that evening the doctor on duty worked with me for an hour before he could release them. It was two weeks before I could straighten my fingers.

My demobilization process didn't take long. First I tore the artillery insignia and sergeant's triangles off my collar, then the star with the hammer and sickle from my fur cap, and threw them away. From one of the deep pockets of his battle dress, Tadek produced a Polish eagle which had been cut out of a tin can. Pinning it to my cap, he said with a solemn face:

"In the absence of *tovarisch* Stalin, I free you, sergeant, from service in the Red Army!"

Then, having clinked glasses, we went out into the street. At a corner stood two soldiers of the NKVD. For an instant the sight of the blue cap produced in me the same old feeling of tension. Noticing it, Tadek pressed my elbow. The tension subsided. As I passed them, talking freely with Tadek, I saw them glance at my uniform, at the eagle on my cap, then look away with bored expressions.

I realized fully for the first time that this was the end of the nightmare. Behind me, under the table in the wineshop, the star with the sickle and hammer was lying on the floor. The bored look in the eyes of the NKVD men was a symbol of a closed chapter in my life.

www.ingramcontent.com/pod-product-compliance
Lightning Source LLC
Chambersburg PA
CBHW072003110526
44592CB00012B/1192